IMAGES
of America

ONIZUKA
AIR FORCE BASE

IMAGES
of America

ONIZUKA
AIR FORCE BASE

Joseph T. Page II

ARCADIA
PUBLISHING

Published by Arcadia Publishing
Charleston, South Carolina

Printed in the United States of America

Library of Congress Control Number: 2019947065

For all general information, please contact Arcadia Publishing:
Telephone 843-853-2070
Fax 843-853-0044
E-mail sales@arcadiapublishing.com
For customer service and orders:
Toll-Free 1-888-313-2665

Visit us on the Internet at www.arcadiapublishing.com

*This book is dedicated to my parents, Joseph
and Kathleen, and my sister Erin.*

The author (left) and his father share a smile while standing on the grounds of the former Onizuka Air Force Base in December 2016. (Author.)

CONTENTS

ACKNOWLEDGMENTS

Foremost, thanks to my parents, Joseph and Kathleen. They are the reason I grew up at Moffett Field and was exposed to this life and location at an impressionable age.

Special thanks to Lorna Onizuka for her supportive reply when I mentioned this project years ago. I hope this book will do justice to the memory of her husband and his legacy.

Shout-out to the NRO's Information Access and Release Team and its declassification efforts.

Col. Andy Wulfestieg (former 21st Space Operations Squadron commander) was instrumental in connecting me with former Onizuka employees and assisting me with the book's scope. Gracious thanks to the anonymous Onizuka employees, whom I've mysteriously named "the Committee," for their unpublished volume on the site's history.

Thanks to Jerry Race and his unexpected email and creativity with concrete crafts and Bill Stubkjaer, Moffett Field Historical Society Museum curator, for letting me browse through its collection of Onizuka newsletters. High-five to Stan Bielecki, Melvin Schuetz, and Steve Olson. These Cold Warriors stepped briefly out of the shadows to assist me. High-five to Dr. Dwayne Day for linking me with Phil Pressel, "Mr. Hexagon."

Fred Taleghani, Randy Rhody, and Ian Abbott provided photographs during and after the demolition of Building 1003. If not for them, a portion of this book's final images would not exist. The connection of Building 1003 to the Manned Orbiting Laboratory program was made by Tom Carr and his son Robert. Tom talked to me for hours about "the Stick" and provided line blueprints for the Advanced Satellite Test Center.

Thanks to Donald "Jay" Prichard, the executive director of the Vandenberg Space and Missile Technology Center, for providing the caricature of Brig. Gen. Bill King. A million thanks to Col. David Arnold (USAF, retired), for writing the seminal work on the Air Force Satellite Control Facility. Each reading of his book gives me greater insight to the complexity of this "living organism" that continues to thrive.

Heartfelt gratitude goes out to Stacia Bannerman at Arcadia Publishing. She has been subjected to my writing continuously over the last eight years.

And finally, thanks to Laura Babcock and the Sunnyvale Historical Society for supporting this project.

Photographs appear courtesy of:

AFHRA	Air Force Historical Research Agency
CIA	Central Intelligence Agency
NRO	National Reconnaissance Office
SAMTEC	Space and Missile Technology Center
USAF	United States Air Force

Introduction

Author's note on naming: Historically speaking, Air Force stations (AFS) are named after locations, not people. On the rare occasion there is an AFS with a memorial name, it is due to downsizing or realignment of the facility. I chose this book's title based on the dedication of the site immediately after Colonel Onizuka's untimely death in 1986. Any perceived slight to former employees about the naming is solely the author's fault.

Right off the intersection of State Route 237 and Mathilda Avenue in Sunnyvale resides a 19-acre plot of land that has remained virtually untouched by the billion-dollar technology companies surrounding it. Through the eyes of its workers and infrastructure, this location has seen many changes over the decades. From fruit groves on dirt roads to glistening glass towers, the land has seen a technological shift.

Onizuka Air Force Station was dedicated in 1986 and named after the first Asian American astronaut, Col. Ellison S. Onizuka. It was renamed Onizuka Air Force Base in 1987. The site, however, had gone by many other names over the decades, some official and some unofficial. The Satellite Test Annex (STA), Satellite Test Center (STC), Air Force Satellite Control Facility (AFSCF), and Consolidated Space Test Center (CSTC) were just a few of the official Air Force designations. The unofficial names were a bit more colorful: the "Stick," "DICE," "Oz," and the one name with staying power that is wrapped in mystery: the "Blue Cube."

Birthplace of Space Reconnaissance

After two pivotal historic events—the surprise attack on Pearl Harbor on December 7, 1941, and the dropping of the atomic bomb on Hiroshima on August 6, 1945—the American psyche was attuned to avoiding the worst possible combination of both events on the North American continent: an atomic sneak attack. Advances in aerial photography, rocketry, and computer technology during World War II helped American scientists and engineers synthesize a solution to this "nuclear Pearl Harbor." The US Army Air Forces–funded Project RAND (adopting its name from a contraction of the term "research and development") released its first report, titled "Preliminary Design of an Experimental World-Circling Spaceship," in May 1946. This report discussed the design, performance, and potential uses of manmade satellites, including an application as a reconnaissance system. In January 1954, Air Research and Development Command (ARDC) used another RAND satellite study, "Project FEEDBACK," to form a proposal for the Advanced Reconnaissance System, also known as Weapon System 117L (WS-117L).

SUNNYVALE AND LOCKHEED MISSILES AND SPACE COMPANY

In 1950, the agricultural community of Sunnyvale, known for the Libby's fruit packing company and the Joshua Hendy Iron Works, had reached a population of just over 9,800. Over the next four years, many industries moved to the city, boosting the population to more than 22,000. The influx of people and industries saw rising taxes placed on farmland, making agriculture unsustainable and encouraging landowners to sell. The Lockheed Aircraft Corporation, originating from Van Nuys, California, created the Lockheed Missile Systems Division, which subsequently became the Lockheed Missiles and Space Company. A new complex, designated as Plant One, was built on 345 acres known locally as the Holthouse property on the perimeter of Naval Air Station (NAS) Moffett Field. Plant One and Lockheed's operations covered 955 acres by 1960. Lockheed considered two critical factors in relocating to Sunnyvale: proximity to institutions of higher learning (such as Stanford University, the University of California–Berkeley, and San Jose State University) and proximity and access to a secure airfield (at NAS Moffett Field).

WEAPON SYSTEM 117L

On October 29, 1956, Lockheed was selected by the Air Force Western Development Division as the primary contractor for WS-117L. The design included many individual subsystems, designated by an alphabet character:

A: Airframe (Agena)
B: Propulsion (Agena)
C: Auxiliary Power Supply (Agena)
D: Guidance and Control (Agena)
E: Visual Reconnaissance (Corona)
F: Ferret (Electromagnetic) Reconnaissance (Signals Intelligence, or SIGINT)
G: Infrared Reconnaissance (Missile Defense Alarm System)
H: Communications System (Satellite Control Network)
I: Data processing
J: Geophysical environment
K: Personnel
L: Biomedical Recovery

Lockheed's WS-117L entry would see the development of the Agena upper stage vehicle and the beginnings of a tracking network (Subsystem H) with a handful of stations around the United States at places like Kaena Point, Hawaii; New Hampshire; Point Mugu, California; and Kodiak, Alaska. Along with the STC, these remote tracking sites (RTS) formed the nucleus of what would later be known as the Air Force Satellite Control Network (AFSCN).

ORIGINS OF THE "STICK"

In a November 14, 1958, memo to the chief of staff of the Air Force, Maj. Gen. Bernard Schriever outlined the offer from Lockheed Aircraft Corporation to transfer 11.4 acres at its Sunnyvale facility to the Air Force without cost. Some of the factors in favor of using the Lockheed offer were proximity to the Lockheed plant where Agena was being built; cost savings from using the orbital determination (ephemeris calculating) computer already present; and simplified management using onsite management and technical personnel. The people were already trained, computers were available, and the land was free. It was a win-win situation for both Lockheed and the Air Force.

While the details for a permanent control station were being planned, Lockheed completed an interim satellite control center in nearby Palo Alto in January 1959. The interim control center would provide command and control support for the February 28 launch of the first Discoverer mission, intercepting 514 seconds of telemetry from the world's first polar-orbiting satellite. The mission would also see the first successful test of the Agena upper stage.

THE NATIONAL RECONNAISSANCE PROGRAM

Starting as a CIA effort under the guise of the Discoverer biomedical satellite project, the Corona program became the world's first photoreconnaissance satellite system. Corona used the Air Force's SM-75 Thor missile and Lockheed's Agena upper stage (and command and control system) to lift a camera, film, and recovery system into orbit. The eventual string of successes from the Corona system—starting with "Lucky 13" (Discoverer XIII)—and other reconnaissance satellites made a strong case for the establishment of the National Reconnaissance Program (NRP). The NRP consisted of all satellite and aircraft reconnaissance projects (including the U-2 and A-12 spy planes), along with photographic projects for intelligence, geodesy, and mapping purposes, with electronic signal collection projects for electronic intelligence and communication intelligence. To oversee the development and operation of NRP systems, the National Reconnaissance Office (NRO) was created, staffed with a mixture of Air Force, CIA, and contractor personnel.

During the early Corona flights, the CIA decided on target coverage (what to photograph), and staff officers at a covert CIA/Lockheed facility in Sunnyvale provided a hard copy of the targeting instructions to STC personnel, who in turn translated the instructions into "vehicle language," commands for the satellite's camera and pointing subsystem. After completing the photographic passes over the target, the STC would receive telemetry from the ground stations so the CIA personnel on hand could verify the health and status of the satellite.

While Corona provided broad-area search capabilities, another satellite, KH-7 Gambit, was designed as a close-look reconnaissance system. While first flights of Gambit produced imagery in the 10-foot to 15-foot resolution range, later Gambit and follow-on KH-8 Gambit-3 flights saw the resolution increase down to better than a foot. During testing for Gambit Mission 4006, launched on March 11, 1964, a Captain Kohlhaas was photographed in the STC parking lot next to his vehicle, becoming the first person to "pose" for a satellite photograph.

By July 1965, combining the Air Force Space Systems Division Deputy for Space Test Operations with the 6594th Aerospace Test Wing created the Air Force Satellite Control Facility, an organization consisting of a headquarters in Los Angeles; a support squadron, Detachment 1, at Sunnyvale; the 6594th Recovery Control Group in Hawaii; and several tracking stations around the world.

After establishment of the AFSCF, the Mission Control Complex (MCC) concept was created to help consolidate control elements for specific satellite programs while maintaining maximum flexibility. Replacement of the CDC-1604 computers with CDC-3600s and later 3800s gave the STC the computing power required for the many supported satellite programs, with appropriate backups in case of failure. More computing power did not equal a more efficient network, so plans to upgrade the data transmission systems were put into place.

Upgraded systems within the NRP also helped guide the development of an Advanced STC (ASTC). The proposed mission of the ASTC was to provide centralized management, direction, and control of the readiness, execution, and evaluation activities of test force elements in support of program test operations. A driving factor behind ASTC's creation was the development of the Manned Orbiting Laboratory (MOL). While the MOL's public story was one of test and development for sustained manned spaceflight, its true purpose was part of the NRP: a manned spy satellite using high-resolution cameras aimed by astronauts inside the spacecraft. ASTC's increased data needs were brought on partly by MOL, allowing it to perform additional missions in an unmanned mode. Cancellation of MOL did not stop construction on the new four-story facility, which was designated Building 1003 but better known by its colloquial name: the Blue Cube.

In 1995, the Base Realignment and Closure (BRAC) commission recommended Onizuka be realigned, with large portions of its AFSCN mission heading to Colorado Springs, Colorado. In 2005, BRAC struck again, this time recommending closure to consolidate satellite command and control operations while reducing excess infrastructure. The recommendation given to the secretary of defense was: "Close Onizuka Air Force Station, CA. Relocate the Air Force Satellite Control Network mission and tenant Defense Information Systems Agency Defense Satellite Communications System mission and equipment to Vandenberg Air Force Base, CA."

The proliferation of NRO mission ground stations and consolidation of Air Force missions at Schriever Air Force Base gave the commission no choice but to move the AFSCN secondary node mission to Vandenberg. Onizuka Air Force Station became a victim of changing times.

One

ELLISON SHOJI ONIZUKA
1946–1986

Ellison Shoji Onizuka, the eldest son of Masamitsu and Mitsue Onizuka, was born on June 24, 1946, in Kona, Hawaii. His parents were merchants who opened a general store in the early 1930s to serve the needs of local farmers. As he grew up in the middle of coffee plantations, Ellison was surrounded by two older sisters, Shirley and Norma, and one younger brother, Claude. In school, Ellison was a determined young man, participating in extracurricular activities—he was on the staff of the school paper, junior class treasurer, and in the National Honor Society, to name a few. His athletic ability also shone through in baseball, basketball, and many intramural sports. As a teenager, he earned the coveted Eagle Scout through exemplary work with the Boy Scouts of America. He graduated from Konawaena High School in 1964 and departed Hawaii for college in Colorado soon after.

Ellison chose the University of Colorado, Boulder, for its aerospace engineering program, earning both a bachelor's and a master's degree in aerospace engineering in 1969. He participated in the university's Air Force ROTC, earning a commission as a second lieutenant, and the distinguished military graduate entered active duty in January 1970. Serving as a flight test engineer and later graduating from the USAF Test Pilot School, Onizuka learned how to judge performance, stability and control, and systems flight testing of aircraft, logging more than 1,700 hours of flying time in various types of aircraft.

Selected as an astronaut candidate in 1978, Onizuka completed a one-year training and evaluation program in 1979. He worked on orbiter test and checkout teams and on the Shuttle Avionics Integration Laboratory team before being selected for his first space flight aboard *Discovery* in 1985, becoming the first Asian American in space. STS-51-C was *Discovery*'s first mission for the Department of Defense, and Onizuka was responsible for the Inertial Upper Stage (IUS). He was selected for his second space mission aboard *Challenger* soon after. After the *Challenger* tragedy, Onizuka's accomplishments were recognized by many memorials, including the renaming of a nondescript Air Force station in Silicon Valley in July 1986.

This official STS-51-L crew photograph of Lt. Col. Ellison S. Onizuka was taken four days before the ill-fated *Challenger* accident. As a spaceflight veteran of STS-51-C, Onizuka's powder-blue NASA flight suit is adorned with the previous mission patch. (NASA.)

ASTRONAUT CANDIDATES
SELECTED JANUARY, 1978

NASA's astronaut candidate class of 1978, unofficially known as the "TFNGs," included six women, three African Americans, and an Asian American (Onizuka, second row from bottom). This was the first new group of astronauts since 1969, hence the acronym. Wanting to introduce diversity into the astronaut corps for the space shuttle era, NASA contracted actress Nichelle Nichols (Lieutenant Uhura on *Star Trek*) to recruit for the agency in the late 1970s. (NASA.)

The crew of mission STS-51-C poses in NASA flight suits. From left to right are (first row) pilot Loren Shriver and commander Thomas K. Mattingly II; (second row) payload specialist Gary E. Payton, mission specialist James F. Buchli, and mission specialist Ellison S. Onizuka. Astronaut Payton's USAF shoulder patch identifies him as a manned spaceflight engineer (MSE), a Department of Defense effort to create a cadre of military astronauts. (NASA.)

The same crew is seen here. From left to right in their military uniforms are (first row) Lt. Col. Loren Shriver (USAF) and Capt. Thomas K. Mattingly II (USN); (second row) Maj. Ellison S. Onizuka (USAF), Lt. Col. James F. Buchli (USMC), and Maj. Gary E. Payton (USAF). STS-51-C was the first dedicated Department of Defense flight in the space shuttle program. Few flight details were acknowledged before, during, or after the mission. (Department of Defense.)

Discovery takes off from Cape Canaveral, Florida, on January 24, 1985, for mission STS-51-C. *Discovery* was the first shuttle orbiter built to lift reconnaissance payloads into orbit. *Discovery* was intended to be stationed at Vandenberg Air Force Base in California for polar-orbiting missions, giving the Air Force a "Blue Shuttle" (military) launch capability. Weighing 6,870 pounds less than *Columbia* or *Challenger*, *Discovery* could lift a larger payload into a polar low earth orbit than the earlier orbiters. During STS-51-C post-mission analysis, NASA engineers discovered the O-rings, required to seal the hot gases inside the combustion chambers of the solid rocket booster while firing, showed heavy charring. According to the Rogers Commission report, this defect, combined with cold weather, was the primary contributor to the explosion of *Challenger* during STS-51-L almost one year later. (NASA.)

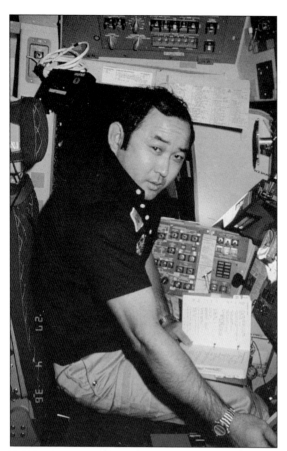

El Onizuka sits in the commander's seat while running a checklist. The flight of STS-51-C lasted three days, one hour, and thirty-three minutes; this was the shortest operational flight and the third-shortest in the entire 135-mission shuttle program. During the flight, NASA would broadcast short messages stating that the flight was "nominal" but nothing more. One could assume the AFSCF in Sunnyvale was busy during the same period. (NASA.)

El Onizuka and Loren Shriver experience weightlessness. The camera provides an artificial sense of what is "up" or "down," but in microgravity the references are meaningless. One interesting detail about STS-51-C was released by NASA in post-flight documents: "The U.S. Air Force Inertial Upper Stage booster was deployed and met the mission objectives." (NASA.)

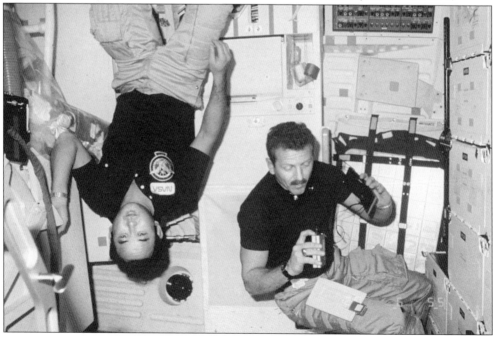

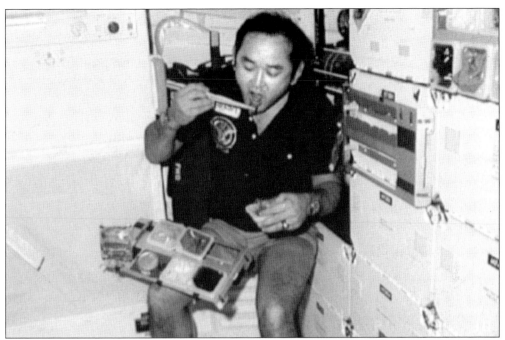

Above, El Onizuka eats a meal with chopsticks in microgravity. The early days of the space shuttle program saw astronauts try new activities in microgravity, such as eating new foods and playing with simple toys, to compare against their earthbound experiences. As a second-generation Japanese American, El Onizuka was extremely proud of his heritage and took opportunities to share his experiences and cultural nuances with the crew and members of the American public. His wearing a *hachimaki*, or stylized headband, with the phrase "Kamikaze" on the side seems to echo stories of Onizuka's sense of humor from his astronaut colleagues. (Both, NASA.)

The logo of STS-51-C, NASA's 15th overall shuttle flight, prominently displays heraldry common within military emblems. The eagle holding arrows in its talons is part of the US Department of Defense logo. The heraldic shield in the center shows a shuttle with a red, white, and blue ribbon streaming from the Earth. The names of the crew are spaced around the patch, with MSE Payton's name at the bottom. (NASA.)

The STS-51-C crew is seen outside the Consolidated Club in late 1984. During this visit to Sunnyvale AFS, the crew followed strict operational security procedures. They filed multiple flight plans to obscure their arrival, told no one of their destination, and drove a junky rental car. When they arrived at their hotel, they were astonished to see "Welcome STS-51-C Astronauts" and their names on the marquee. (Onizuka Alumni.)

This final portrait of the STS-51-L crew was taken in November 1985. From left to right are (first row) Michael J. Smith, Francis R. "Dick" Scobee, and Dr. Ronald E. McNair; (second row) Ellison S. Onizuka, Sharon Christa McAuliffe, Gregory Jarvis, and Dr. Judith A. Resnik. (NASA.)

The patch design for STS-51-L included items reflecting specific aspects of the planned mission. Within the canton, the blue field containing seven white stars, is a gray outline of a comet. Part of *Challenger*'s mission was to observe Halley's Comet for six days after deployment of its primary payload. The Teacher in Space Program is recognized by a small red apple between payload specialists Christa McAuliffe and Gregory Jarvis's names. (NASA.)

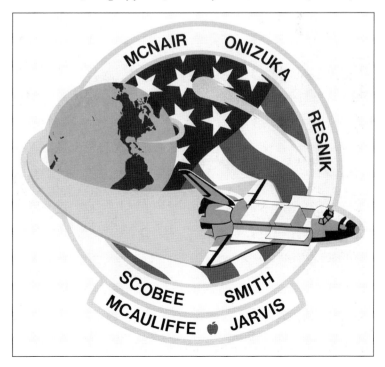

Nineteen days before launch, the STS-51-L crew posed for this photograph inside the White Room atop Launch Complex 39B. From left to right are teacher in space and payload specialist Christa McAuliffe; payload specialist Gregory Jarvis; mission specialist Judith A. Resnik; mission commander Francis R. "Dick" Scobee; mission specialist Ronald E. McNair; pilot Mike J. Smith; and mission specialist Ellison S. Onizuka. Note the impression on Onizuka's face made by the tight-fitting helmet. (NASA.)

Onizuka (far left) and four members of the STS-51-L prime crew listen during a training session at the Johnson Space Center in Houston, Texas. The sixth individual (far right) is Barbara Morgan, the back-up payload specialist to Christa McAuliffe and second Teacher in Space Program selectee. Twelve years after the *Challenger* accident, Morgan joined the astronaut corps as a mission specialist, finally flying into space on STS-118 in August 2007. (NASA.)

Mated to its Inertial Upper Stage, the fifth Tracking and Data Relay Satellite (TDRS) is released by STS-43 in August 1991. The primary payload aboard *Challenger*'s last flight was the second TDRS, poised to supplement TDRS-1, launched aboard *Challenger* in 1983. During the TDRS-1 deployment, failure of the second stage placed the satellite in an incorrect orbit. Curiously, Onizuka was the only STS-51-L crewmember with previous experience with an IUS. (NASA.)

Ellison Onizuka was laid to rest on June 2, 1986, during a Buddhist ceremony at the National Memorial Cemetery of the Pacific in Honolulu, Hawaii. Over 700 mourners attended the service, including his wife, Lorna; two daughters, Janelle and Darien; mother, Mitsue Onizuka; and STS-51-C crewmate Col. Loren Shriver. Norman Sakata, Onizuka's scoutmaster, spoke reverent words about El's influence: "He was an inspiration not only to the young, but to all of us." (William Balco Sr.)

A well-traveled soccer ball floats inside the cupola on the International Space Station. This ball was signed by the 1986 Clear Lake High School (Texas) women's soccer team, including Onizuka's eldest daughter, Janelle. Colonel Onizuka had intended to take the ball into space on STS-51-L. The ball was retrieved from the wreckage of *Challenger*, and in 2016, Expedition 50 commander Shane Kimbrough took the ball to the station for 173 days in space. (Shane Kimbrough.)

Two

BIRTHPLACE OF SPACE RECONNAISSANCE 1959–1969

Lockheed completed an interim satellite control center in Palo Alto in January 1959. During the launch of Discoverer I on February 28, the interim facility received a total of 514 seconds of telemetry from the world's first polar-orbiting satellite. That mission was the first to use a Thor as a space booster, and it also represented the first successful flight test of Lockheed's Agena-A upper stage vehicle, designed for orbiting US satellite systems. On April 6, 1959, Headquarters Air Research and Development Command established the first military unit to be charged with conducting military satellite operations, the 6594th Test Wing in Palo Alto. The operations of the test wing moved to Sunnyvale into another interim control center on the Lockheed campus in March 1960, awaiting the opening of the Satellite Test Center inside Building 1001.

Changes for the facility proceeded at a rapid pace in the next few years. The control network grew from the Sunnyvale command center and two remote stations supporting WS-117L (1958) to a five-station network for Discoverer (1960) and to a seven-station network for Department of Defense/NRO satellite programs in 1961. Mission requirements for the network by 1962 included:

- A single satellite being tracked by any one of seven ground stations at any one time, with realignment for the next satellite pass within 30 minutes
- Four satellites being commanded/controlled simultaneously at the Satellite Test Center
- One vehicle being recovered (re-entered) during a 24-hour timeframe
- Simultaneous launch support and single satellite pass from Vandenberg Tracking Station.

Integration of the Gambit and photographic satellites into the National Reconnaissance Program, along with the Earpop signals intelligence payloads in the mid-1960s, increased the workload of the STC and its remote tracking stations. These reconnaissance missions were run in plain sight of other Department of Defense missions such as the Space Test Program missions. Engineering lessons learned from these early missions helped during the creation of a satellite Command and Control standard, the Space-Ground Link Subsystem, for future missions. Upgrades to the Satellite Control Facility were planned throughout the end of the decade, including a large military construction project for the Advanced Satellite Test Center.

After 12 failures, "Lucky 13" became the first successful Discoverer mission (Agena 1057/Thor 231), lifting off from Vandenberg Air Force Base on August 10, 1960. By this time, satellite control operations had moved from Palo Alto to a temporary "Satellite Test Annex" in Sunnyvale; the Air Force would later occupy Building 1001 in February 1961. The success of Discoverer XIII paved the way for the first successful Corona camera launch a week later. (SAMTEC.)

A diagram of the Discoverer recovery capsule shows its major subsystems. Note the identification of a "bio pac" (biomedical package) and animal cabin inside the capsule. The Discoverer cover story outlined the program's purpose as biological space research. Some of the aerial crews from the 6593rd Test Squadron thought the recovery capsule contained monkeys, mice, roaches, or other biological specimens. (USAF.)

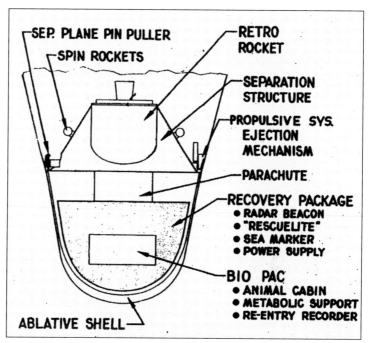

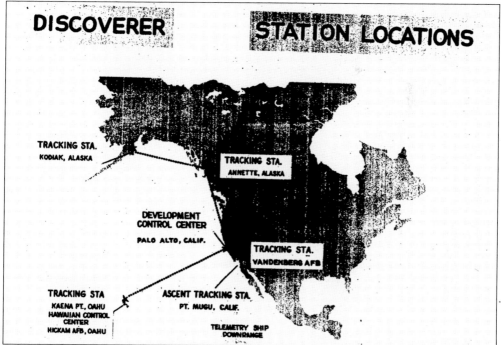

This early map of the Discoverer tracking system identifies the Interim Development Control Center (IDCC) in Palo Alto and tracking stations at Kodiak and Annette Island, Alaska; Kaena Point, Hawaii; and Vandenberg Air Force Base, California. The Hawaiian Control Center directed the specialized C-119J aircraft during satellite recovery operations. In order to extend the reach of the command and control network, missile range instrumentation ships from Point Mugu, California, were also used. (USAF.)

An overhead view of the future home of Sunnyvale AFS shows the Lockheed Missiles and Space Company complex (known as "Plant One"), the intersection of Mathilda Avenue and Highway 237, and the lone operations building, later renamed Building 1001. Near the bottom center of the photograph is Highway 101, complete with a line of traffic and a four-way stop. (Onizuka Alumni.)

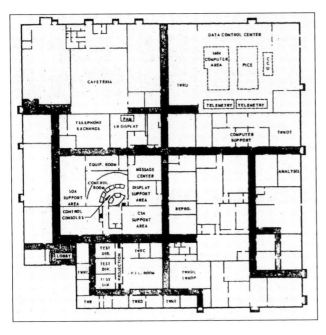

This diagram of Building 1001 shows the delineation of operations (control room and consoles), offices, and the data center. This configuration shows the first construction expansion from the original building, which added over 45,600 square feet. A second expansion in 1963 added 4,200 square feet for communications, crypto, and electro-mechanical spaces. The elliptical control center can be seen near center left. (USAF.)

AGENA B:
America's off-the-shelf satellite

Down the production line at Satellite Center, U.S.A.* move the Agena B satellites now being used in many of the nation's major space programs. The Agena B proved its reliability in the trail-blazing Discoverer program. It is easily adaptable to a wide range of payloads and missions. And its powerful start-and-stop rocket engine can maneuver it precisely to any orbit. The Agena B is built by Lockheed. Major subcontractors: General Electric, Bell Aerospace, and Philco.

LOCKHEED
*MISSILES & SPACE DIVISION • SUNNYVALE, CALIFORNIA

An early promotional advertisement of Lockheed's Agena upper stage shows the propellent and engine rear section attached to a generic payload fairing. After the fairing was jettisoned, reconnaissance payloads rarely showed such aerodynamic curves; portions of photoreconnaissance satellites had bulbous noses (recovery capsules), while SIGINT satellites (such as the P-11 and P-989) had helical and parabolic antennas sticking off the satellite bus for better reception of radio signals. (Lockheed.)

Lockheed contractors inspect the forward auxiliary rack (left) on the Agena-A and the aft rack (right) that would propel Discoverer XIII into orbit. The Agena-A provided an orbital insertion capability along with limited battery life. Later models, such as the Agena-B and the standardized Agena-D, provided additional capability to reconnaissance satellite programs, including engine restart capability and extended battery life. (SAMTEC.)

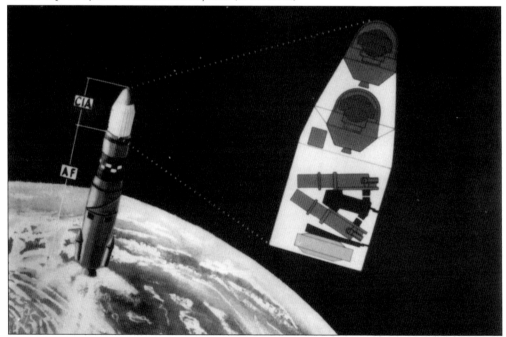

This cut-away view of a Thor/Agena shows the internal configuration of a KH-4B Corona satellite with its cameras, two film reels, and corresponding recovery capsules. This simplified diagram shows the split of Thor/Agena responsibility between the CIA (capsule and camera) and the Air Force (launcher). The STC used Lockheed contractors with CIA and Air Force personnel to control the integrated Agena satellite via the remote tracking stations around the world. (NRO.)

An internal schematic of the KH-4B, the final version of the Corona satellite, shows the J-3 system and two recovery capsules (top). The KH-4A and KH-4B variants carried a film payload of 160 pounds, doubled from the KH-4's 80-pound reels. The two capsules also provided flexibility in targeting, allowing the first to image high-priority targets and be returned soonest while the second remained on-orbit. (NRO.)

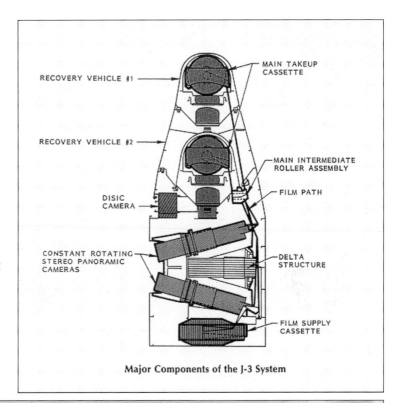

Major Components of the J-3 System

The entrance to Building 1001 at the Satellite Test Center is pictured around 1960. At the time, the Air Force's Air Research and Development Command (ARDC) controlled weapon system research, development, and testing through procurement and production, since many early space systems were ostensibly operational prototypes. Organizational changes saw ARDC renamed Air Force Systems Command in April 1961, with the STC (and later AFSCF) organized within the hierarchy. (USAF.)

Lt. Col. Charles "Moose" Mathison (right), commander of the 6594th Test Wing, sits at a control console, while a contractor assists. The early days of satellite operations at the STC saw the 6594th Test Wing and Lockheed conduct satellite operations with joint responsibility, though contractors operated the mission equipment. In 1960, Air Force members at the STC were outnumbered by contractors two to one. (USAF.)

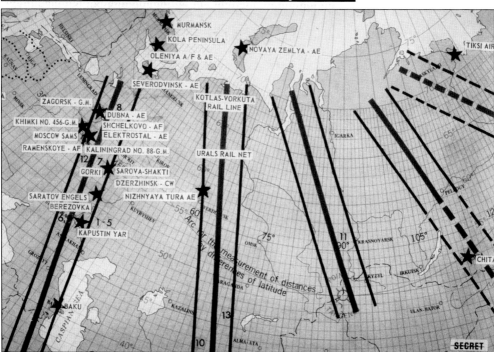

This orbit plot for an unidentified Corona flight shows the notional path (center line), with the side lines representing the maximum extent of photographic coverage on either side. The stars denote photographic target areas. The first successful Corona flight acquired 3,000 feet of film covering more than 1.65 million square miles of Soviet territory—more overhead photographic coverage of the Soviet Union than all the U-2 flights to that date. (NRO.)

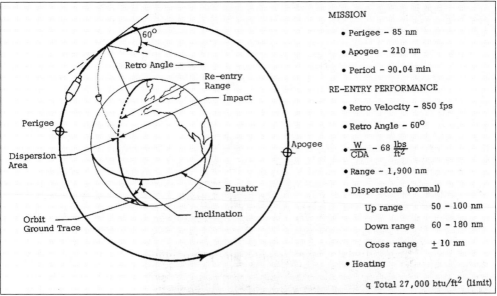

MISSION

- Perigee – 85 nm
- Apogee – 210 nm
- Period – 90.04 min

RE-ENTRY PERFORMANCE

- Retro Velocity – 850 fps
- Retro Angle – 60°
- $\dfrac{W}{CDA}$ – 68 $\dfrac{lbs}{ft^2}$
- Range – 1,900 nm
- Dispersions (normal)

Up range	50 – 100 nm
Down range	60 – 180 nm
Cross range	± 10 nm

- Heating

q Total 27,000 btu/ft^2 (limit)

The orbital mechanics of satellite re-entry and recovery were computed on one of two CDC-1604 computer systems inside the STC. If the re-entry angles were not precise, the capsule could miss the drop zone near Hawaii. This profile would be used by future Corona satellites, as well as Gambit and Hexagon re-entry vehicles, until the early 1980s. (CIA.)

A contractor talks with a military operator sitting in the test director chair inside Building 1001. Test directors maintained expertise on all onboard systems, while test controllers were the operators who actually sent the commands to the satellite. The sinusoidal (wavy) lines on the board are orbital plots of a spacecraft being tracked by the STC. The phase shift of the line is due to the earth's rotation. (USAF.)

The control room inside Building 1001 is seen here. The command consoles, both in Sunnyvale and at the remote tracking stations, were monitored by trained personnel (military or contractor) during the length of the mission. In the early days of Corona, this would equal a few hours on a console. With increasing numbers of reconnaissance missions and longer satellite lifespans, control was exercised with 24-hour support. (USAF.)

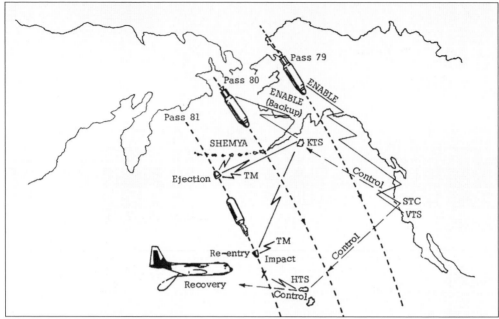

This diagram shows the intricate operation of satellite recovery. Commands to initiate the recovery sequence came from the STC and were transmitted by Vandenberg (VTS) on pass 79. If the command was not received, Kodiak (KTS) would reinitiate the enable command on pass 80. Pass 81 would have Kodiak send the ejection command, while Kaena Point (HTS) would track and send recovery forces via air and sea. (CIA.)

Around 1959, Kodiak was a hodge-podge of buildings, World War II–era antennae, and adventurous contractors. A VERLORT (VERtical LOng-Range Tracking) three-pulse tracking and commanding radar sent the satellite commands during passes. The only offsite communications were sent via 100-word-per-minute teletype and telephone. (USAF.)

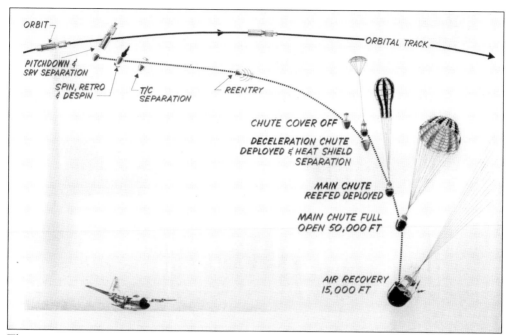

The recovery sequence was initiated by commands from either Vandenberg or Kodiak tracking stations directing the satellite to reorient itself for re-entry capsule ejection. After the heat shield and parachute cover were jettisoned, the main parachute would deploy at around 50,000 feet to allow time for recovery forces to arrive and attempt capture. (CIA.)

A C-119J captures a re-entry capsule in midair. Two 38-foot poles were arranged in a V shape with a double loop of nylon cable and special grappling hooks stretched across the 15-foot gap between the poles. The recovery system absorbed the initial shock of capturing a 300-pound package before it was reeled inside. Used recovery hooks showed prongs deformed from the instantaneous heat produced by the hook ripping across nylon parachute lines. (NRO.)

Aerial recovery loadmasters aboard a C-119J retrieve a Corona re-entry capsule that was captured in midair. Each man wore a standard-issue parachute in case turbulence knocked him out of the aircraft. The cable at lower right center led to the retraction winch, which pulled the capsule inside the aircraft. After capture, the capsule was sealed in a drum for transport to Eastman Kodak facilities in Rochester, New York. (USAF.)

A Navy helicopter from the USNS *Haiti Victory* pulls the Discoverer XIII re-entry vehicle out of the Pacific Ocean on August 11, 1960. If airborne recovery failed, sea recovery was the backup method, using frogmen to secure the capsule and a helicopter to lift it. If friendly forces were unable to retrieve the re-entry capsule in a reasonable amount of time (up to 48 hours), a small salt plug would slowly dissolve and fill the container with water, sinking it to the ocean depths. (NRO.)

Lieutenant Colonel Mathison (far left) smokes a stogie as he operates monitoring equipment inside the STC. Mathison had been hand-picked by Lt. Gen. Bernard Schriever to supervise contractors constructing Thor launch pads at Cape Canaveral. He was later selected as the commander of the 6594th Test Wing at Sunnyvale during the early days of Corona. He finished his career as the commander of Kirtland Air Force Base in New Mexico in 1973. (USAF.)

The Discoverer XIII capsule is delivered by Lieutenant Colonel Mathison to Lieutenant General Schriever and Air Force chief of staff Gen. Thomas D. White on August 13, 1960, at Andrews Air Force Base, Maryland. According to flight crew interviews, during the flight from Hawaii to Moffett Field, Mathison opened the drum containing the space capsule. The trip to Lockheed in Sunnyvale was to remove the SOCTOP electronic intelligence payload. (USAF.)

Pres. Dwight Eisenhower (far left) inspects an American flag from the Discoverer XIII mission as Air Force chief of staff Gen. Thomas D. White and Lieutenant Colonel Mathison look on. While the public presentation celebrated the Discoverer program's "peaceful use of space," the Agena upper stage carried a SIGINT payload called SOCTOP in its aft rack that was removed by Lockheed engineers after recovery. (NRO.)

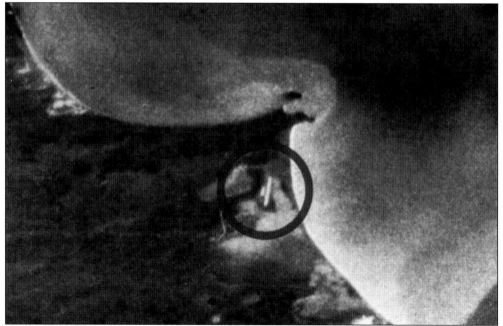

The first image taken from a satellite-based camera was of the airfield at Mys Shmidta, USSR. The circle shows the runway and parking apron at the airfield at a resolution of around 40 feet from the first-generation Corona camera. While the image was too poor to provide any useful intelligence, the first mission proved the concept of space-based photography for US policymakers. Less than 10 years later, photographic resolution would be down to better than a foot. (NRO.)

The master control room inside the STC shows Lockheed test director Eugene Crowther (center right, standing) and test controller Maj. Joseph P. O'Toole (center left, sitting) during the final seconds before a Discoverer launch in 1961. The others in the foreground represent a mix of military personnel and contractors. (USAF.)

The controller stations inside the STC around September 1963 show consoles overflowing with information regarding communications with tracking stations, plotting boards with satellite re-entry information, and screens displaying current weather conditions. The amount of data present and the reach-back capability to request more information via voice or phone were hallmarks of satellite operations at the STC. (USAF.)

STATION	GEODETIC COORDINATES	
	LATITUDE	LONGITUDE
BARKING SANDS PMR FACILITY	22°01'N	159°47'W
CHRISTMAS ISLAND PMR FACILITY	1°58'N	157°48'W
HAWAII TRACKING STATION	21°34'N	158°18'W
KODIAK TRACKING STATION	57°34'N	152°10'W
NEW BOSTON TRACKING STATION	42°56'N	71°37'W
PT. MUGU TRACKING STATION	34°6'N	119°17'W
SOUTH POINT PMR FACILITY	18°56'N	155°41'W
TERN ISLAND PMR FACILITY	23°32'N	166°17'W
THULE, PROJECT SPACE TRACK, TRACKING STATION	76°33'N	68°50'W
WOOMERA SATELLITE TRACKING STATION	31°12'S	137°06'E

A screen shot inside the Building 1001 control center shows the latitude and longitude of the remote tracking locations around the world, along with their voice call-signs. Many of these locations

STATION CALL SIGNS

STATION	VOICE	TTY	SSB
ASCENSION ISLAND AMR FACILITY		YOASC	
BARKING SANDS	SANDY	BARK	
CAPE CANAVERAL AMR I FACILITY		YOCAN	
CHRISTMAS			CHRISTMAS
DOWN RANGE T/M SHIP	HORSE	HORSE	SLEIGH BELL BRAVO
HAWAII CONTROL CENTER	HICK	HICK	HARVEST MOON
HAWAIIAN TRACKING STATION	HULA	HAWA	HULA
KODIAK TRACKING STATION	KODI	KODI	SLEIGH BELL DELTA
NEW BOSTON TRACKING STATION	BOSS	NEWB	
PT. MUGU TRACKING STATION	MUGU	MUGU	BEER FOAM I ALPHA
SATELLITE TEST CENTER	DICE	STC	SLEIGH BELL ALPHA
SOUTH POINT PMR FACILITY	POINTER	SOPT	
TERN ISLAND PMR FACILITY	TERN	TERN	TERN
THULE PROJECT SPACE TRACK TRACKING STATION		THULE	
WOOMERA SATELLITE TRACKING STATION		WOOM	

were integrated into the AFSCN. Other sites remained separate, belonging to other services' space programs or NASA. Note the STC's call-sign of DICE, a corruption of IDCC. (USAF.)

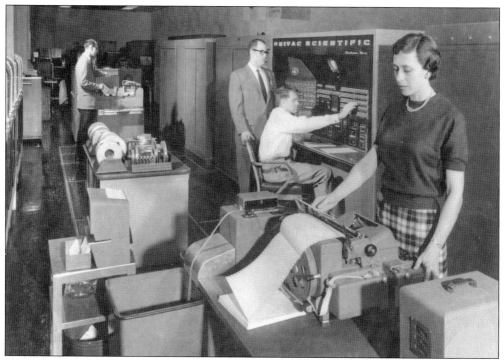

Lockheed personnel operate a Univac Scientific computer system in support of Discoverer computer operations in 1959. (USAF.)

Three personnel confer on system operations in 1961. To get the satellite program operational as soon as possible, off-the-shelf equipment was used as often possible. (USAF.)

Console operators sit inside a mission control complex inside the STC. Teletype messages from the STC forwarded launch and support schedules to the remote tracking stations, with distant operators passing telemetry back to Sunnyvale. Once the MCC concept was implemented, individual satellite passes could be provided to the corresponding control center. (USAF.)

Kodiak was the first place on American soil to pick up the satellite's beacon when traveling a north-south polar orbiting track. The site included two antennae: one for transiting commands to the satellite and a second to receive telemetry. A large X-Y plotter (center, rear) holding preprinted maps tracked the flight's orbit as information was received. (USAF.)

A radome is constructed over the 60-foot Telemetry and Data Receiving Antenna at the New Hampshire tracking station in 1960. The site would not receive certification for operations until July 15, 1961; prior to this, the station was limited to passive tracking during the Discoverer XIII mission. (USAF.)

A quad-helix UHF antenna is pictured at the Kaena Point tracking station. The quad- and tri-helix antennas at the early tracking stations received telemetry from the satellites. This data was recorded and sent back to Sunnyvale via 1,200-bit-per-second lines or magnetic disks flown to California by military courier. (USAF.)

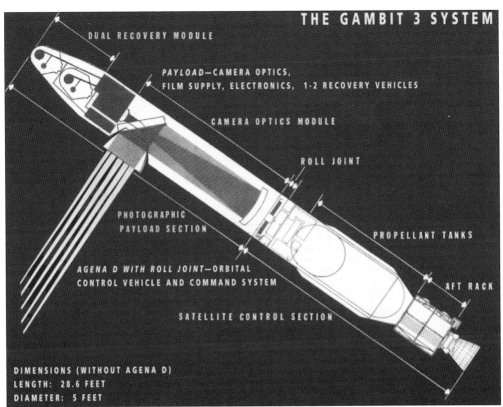

DUAL RECOVERY MODULE

PAYLOAD—CAMERA OPTICS, FILM SUPPLY, ELECTRONICS, 1-2 RECOVERY VEHICLES

CAMERA OPTICS MODULE

ROLL JOINT

PHOTOGRAPHIC PAYLOAD SECTION

PROPELLANT TANKS

AGENA D WITH ROLL JOINT—ORBITAL CONTROL VEHICLE AND COMMAND SYSTEM

AFT RACK

SATELLITE CONTROL SECTION

DIMENSIONS (WITHOUT AGENA D)
LENGTH: 28.6 FEET
DIAMETER: 5 FEET

This side view of a KH-8 satellite shows the optical path from the incoming light (target) reflected against a large mirror and then projected onto the film. Early Gambit flights kept the Agena-D upper stage attached in "hitch-up mode," while later ones discarded the Agena in favor of pointing with the Orbital Control Vehicle. The Gambit series produced extremely high-resolution photographs, the limits of which are still classified decades later. (NRO.)

The second KH-7 Gambit-1 satellite rode aboard an Atlas/Agena-D booster on September 6, 1963. Upgrades to the remote tracking stations allowed multiple-satellite contacts, mirroring the NRP's strategy of launching both Corona and Gambit-1 flights. Within the STC, a Multi-Operations Section was organized with an Orbit Planning Branch directed to create processes and procedures for multi-satellite operations. (SAMTEC.)

TOP SECRET
(CLASSIFICATION)

3|2

EYE-0149-62
Copy 1

(b)(3)

Handle Via Indicated Controls

..B.Y.E.M.A.N..

....

....

....

....

....

WARNING

This document contains information affecting the national security of the United States within the mean-
ing of the espionage laws U. S. Code Title 18, Sections 793 and 794. The law prohibits its transmission or the revelation of its contents in any manner to an unauthorized person, as well as its use in any
manner prejudicial to the safety or interest of the United States or for the benefit of any foreign government to the detriment of the United States. It is to be seen only by U. S. personnel especially indoc-
trinated and authorized to receive information in the designated control channels. Its security must be
maintained in accordance with regulations pertaining to the designated controls.

This document contains information referring to Projects:

..IDEALIST.....   CORONA.....   ARGON.....   OXCART.....   ....

TOP SECRET
(CLASSIFICATION)

A copy of a Byeman Control System coversheet from the early 1960s shows names of reconnaissance programs mentioned within. A "byeman" is someone who works underground. According to NRO documents, when the CIA security officer chose the name for the new security system, he was unaware of the definition or connotation to the "black world" (classified) projects that would be held within the control system. (NRO.)

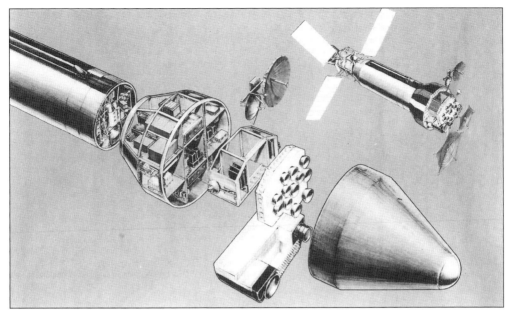

One of the earliest SIGINT satellite programs run out of Sunnyvale was Program 770, under the codeword Earpop. The program had numerous variations, depending on the targeted radio frequencies and antenna design. The payload's integration with an Agena upper stage once again assured Lockheed's expertise would be heavily relied upon during the satellite missions. (NRO.)

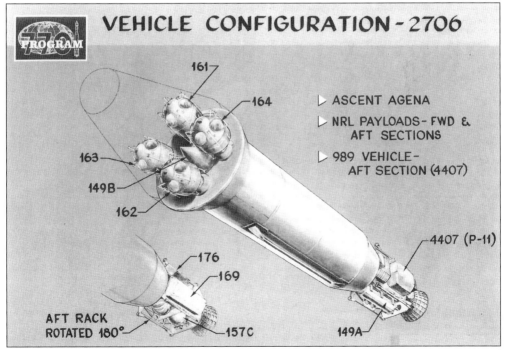

VEHICLE CONFIGURATION - 2706

161
164
163
149B
162
176
169
AFT RACK
ROTATED 180°
157C
4407 (P-11)
149A

▷ ASCENT AGENA
▷ NRL PAYLOADS - FWD & AFT SECTIONS
▷ 989 VEHICLE - AFT SECTION (4407)

This configuration diagram for Mission 2706 (Poppy) shows how early launches were packed with SIGINT equipment. Poppy satellites were held under the shroud, while the aft rack held antennae and equipment from the P-11 program. These programs were only recently declassified, even though their launches were in the early to mid-1960s. (NRO.)

Imagery from Gambit, such as this 1966 photograph of the US Capitol, kept increasing in capability and resolution, necessitating more work from the STC. Missions from both Corona and Gambit were used in a "spotter/shooter" pair, with the KH-4A/B cameras performing wide-area search and the KH-7 pulling in high-resolution images of targets of interest. (NRO.)

Brig. Gen. William G. King Jr. began his special projects association in the late 1950s with the Samos Program Office, which became the NRO Office of Special Projects in 1961. He commanded the AFSCF as a colonel from 1966 to 1967 and later became the director of NRO's Program A before his retirement. (USAF.)

While in Sunnyvale, an oft-heard quote from Brigadier General King was "I could run this operation with a phone booth and a roll of nickels." At his retirement ceremony in 1971, King was presented with this caricature of him pulling a phone booth. Based on the nickel reference, some of his subordinates wondered when King had last made a call at a payphone, hence the roll of dimes seen here. (SAMTEC.)

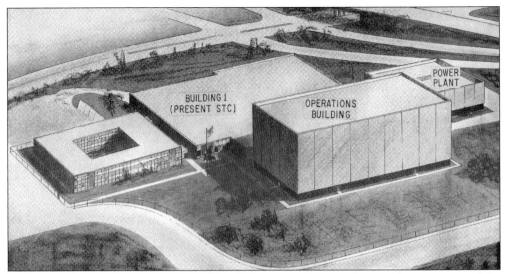

From the Aerospace Corporation's Technical Operations Report, this artist's conception of the Advanced STC shows its companion power plant (Building 1004) and connection to the "present STC" inside Building 1001. Aerospace planners could not predict the demise of the MOL program, nor would they accurately predict the increase in national security space missions throughout the 1970s. (Tom Carr.)

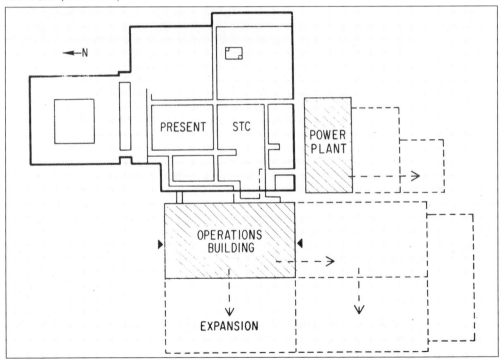

The plan of expansion beyond ASTC is shown here. While the area around Building 1003 did eventually expand to include additional MCCs, it is unknown from this drawing and the corresponding report if the three additional expansion slots were meant for multistory buildings (duplicating the Blue Cube) or something smaller, similar to what was eventually built around Building 1003. (Tom Carr.)

Three

HIP TO BE CUBED
1969–2009

Completion of Building 1003, colloquially known as the Blue Cube, was a transitional period from the first phase of reconnaissance and satellite command and control to the second, with higher resolution satellites. The building's raison d'etre, the MOL program (also known as KH-10 Dorian) was cancelled in July 1969 due to the ballooning system cost and the Vietnam War; the death knell for MOL was the development of a near-real-time electro-optical satellite system.

Termination of the MOL program eliminated the requirement for the Advanced Data System and the MOL portion of the ASTC. However, the US Air Force evaluated current and future requirements for satellite programs, determining that construction should continue. The 104-foot-high windowless building was painted Air Force blue, giving rise to its nickname, "the Big Blue Cube."

In July 1971, the Satellite Test Annex was renamed Sunnyvale Air Force Station. Official use of the acronym STA was terminated; however, the facility's original nickname of the "Stick" was used locally to distinguish between the operational command center and the facility as a whole. Almost a year later, on May 25, 1972, Sunnyvale AFS supported the final launch of the Corona program from the Satellite Control Room in Building 1001, marking the end of the first-generation photoreconnaissance satellite program.

In the 1970s, satellite reconnaissance work was booming. The first KH-9 Hexagon mission was launched in 1971, followed in 1976 by a near-real-time electro-optical satellite designated KH-11 Kennen. The civilian space program followed with a Space Transportation System, a space "truck" that would lift civil and commercial payloads into orbit. As part of an agreement with NASA, the Department of Defense agreed to transition many satellites off of expendable booster rockets and onto the shuttle. Military payloads would continue to be operated from Sunnyvale, while the Blue Cube began supporting space shuttle missions in 1981.

12 JUL 68

At the time construction started on the Advanced STC, it was destined to house command and control for the MOL program. The Air Force purchased an additional 8.2 acres of land from Lockheed, which was finalized in August 1968. The building would house communications and cryptographic equipment, along with large mainframe computers with an extensive tape library. (Tom Carr.)

In 1963, Secretary of Defense Robert S. McNamara announced the cancellation of the X-20 DynaSoar project in favor of the Manned Orbiting Laboratory (MOL, above). The Department of Defense press release stated that the MOL would "increase the Defense Department's effort to determine the military usefulness of man in space." In reality, the MOL was a cover for a manned reconnaissance satellite, designated KH-10 Dorian (below). A team of two military astronauts would ride a Gemini-B capsule atop the MOL for missions lasting upwards of 30 days. This image shows the pressurized lab module, which would house the astronauts, and the unpressurized mission module containing the mirrors and camera. An early MOL concept included an unmanned mission configuration that would be remotely controlled from Sunnyvale. (Both, NRO.)

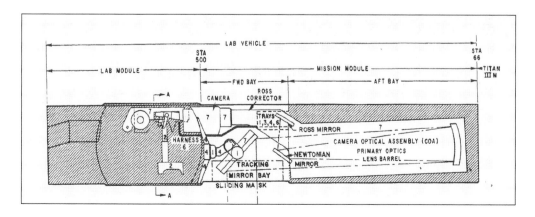

In September 1968, construction crews started to attach the iconic concrete panels to the outside of Building 1003's frame. Each panel was made of three-inch-thick concrete and was attached to the steel frame at each corner. The panels were created onsite with concrete pours into premeasured molds. (Onizuka Alumni.)

This aerial view of Sunnyvale AFS in the late 1970s shows the changes made to the facility through its first decade and a half. While the nomenclature of Satellite Test Center had been officially eliminated in July 1960, it remained in daily use and within formal publications. The large three-foot-high letters on Building 1003 stating "Satellite Test Center" did not help dispel the confusion. (USAF.)

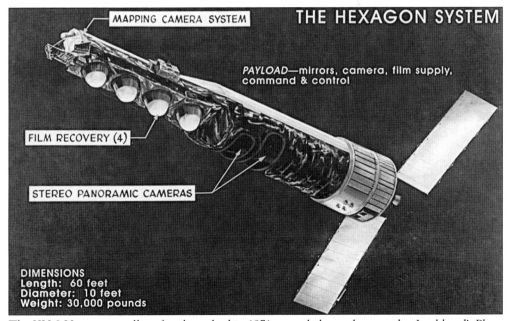

The KH-9 Hexagon satellite, first launched in 1971, was a behemoth created at Lockheed's Plant One. Advances in space communications technology, along with the Tunity command and control software, pushed Hexagon to its maximum capability as a space-borne reconnaissance system. (NRO.)

13 NOV 68

Building 1003 is seen here in November 1968, with panel emplacement nearing completion. The panels were designed to break away from the structure during an earthquake of magnitude 8.0, saving the structure's internal integrity. Note the distance of Lockheed/Innovation Way from the building's perimeter. (Onizuka Alumni.)

The logo of the AFSCF adopted heraldry elements from the 6594th Aerospace Test Wing, retaining the bird of prey orbiting the earth. Air Force adoption of the olive drab fatigue uniform saw the creation of a subdued version of this patch for everyday wear. (AFHRA.)

While media sources at the time stated the AFSCF was a high-security facility with very little reported about it, this screen shot of the Sunnyvale Air Force Station sign was taken from an unclassified Air Force introductory film detailing the operations of the center in the early 1980s. Plenty of psychedelic colors and animated graphics described how the AFSCF controlled US satellites in orbit. (USAF.)

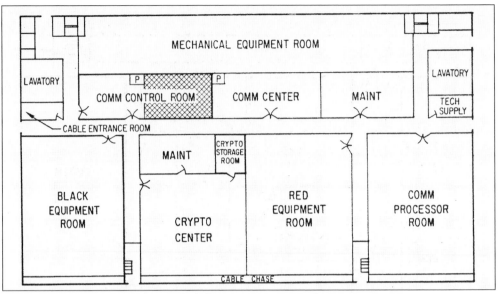

Building 1003's first-floor plan included a crypto center and communications processor room, providing the ability to send encrypted messages to a variety of locations around the world. In many communication centers, "red" (encrypted) and "black" (unencrypted) equipment work conceptually side-by-side to give commanders operational flexibility during missions. In reality, the black and red systems required separation due to electromagnetic interference requirements. (Tom Carr.)

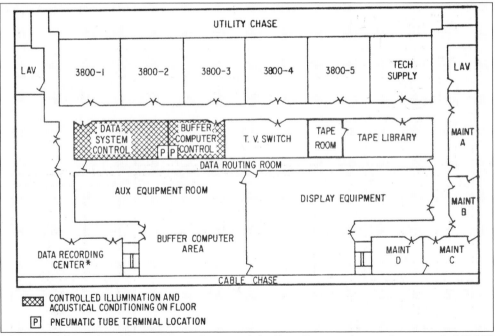

The second floor of Building 1003 was designated the Data Systems Area and consisted of 27,000 square feet of technical floor space. The bulk of the data handling and processing for the AFSCF was accomplished here by five secured computer rooms containing the CDC-3800 computers. (Tom Carr.)

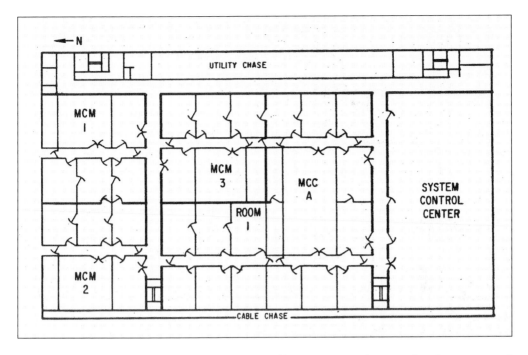

The third (above) and fourth (below) floors of Building 1003 were designated as the operational floors containing the Mission Control Complexes. A modular floorplan was developed to meet the varied requirements for satellite (family) mission control. A typical MCC consisted of three operations rooms with three technical support areas arranged in a 60-by-60-foot square configuration. (Both, Tom Carr.)

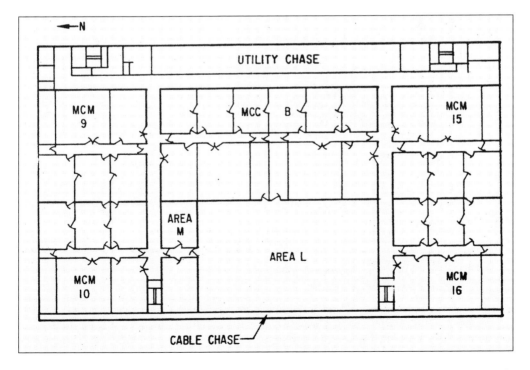

Building 1004 contained the power plant, providing uninterrupted electricity to the entire base. As energy prices spiked during the 1970s, uninterrupted electricity became a necessity for the installation's support to critical satellite missions. Building 1004 held gas turbine generators powering the entire base. Connections to California's Pacific Gas & Electric (PG&E) network were severed, rendering it energy independent. (USAF.)

Personnel monitor Building 1004's power plant control equipment. The power plant was staffed by contractors from Solar, a division of the International Harvester Company, to provide power to the entire installation. One power line to PG&E was maintained through a Lockheed substation to provide backup power capability if needed. (Onizuka Alumni.)

One of the centers inside Building 1003 shows the sheer number of people required to run the computing equipment. Early network configurations had comparable systems placed at select RTSes to speed up processing of telemetry and payload data before sending to the STC. More computers and better telecommunications lines in Sunnyvale reduced the need for heavy computing power at the tracking stations. (Both, USAF.)

The Sun East and Sun West dishes supported connectivity between satellites in geostationary orbit and customers ranging from presidential communications to missile warning satellites. The dishes were capable of simultaneous transmission and reception of 24 and 32 links respectively. The terminals held a mission availability rate of over 99.9 percent due to automated switchover redundancies. (Both, Onizuka Alumni.)

A contractor computer operator talks to a tracking station while monitoring the satellite contacts. While Air Force leadership desired a full military-run operation at the AFSCF, issues with personnel training and transfers solidified the position of having a dedicated contractor staff working in conjunction with the military. Some of the "graybeards" would remain at Sunnyvale for decades after transitioning out of active military service into a contractor job. (USAF.)

An airman attempts to swap out a disk platter from a Univac 1531's CDC-854 disk drive unit. These disk drive units could read and record data written magnetically on six removable disk pack units. The total storage capacity per disk is 49,152,000 bits, or roughly six gigabytes of information, with a data transfer rate of 1.25 megabits per second. (USAF.)

This Hexagon image from Mission 1219 shows Moffett Field with Hangar Two and Hangar Three near the center. Scores of the US Navy's main anti-submarine aircraft, the P-3 Orion, are seen on the ramp. Immediately east (left) of the photograph's coverage (off-image) are Sunnyvale AFS and the Lockheed complex. (Phil Pressel.)

Considered state-of-the-art at the time and a precursor to touch-screen technology, a light pen allowed an operator to select or deselect information on the screens. This interface was usually faster than operators searching through menus via keyboard. (USAF.)

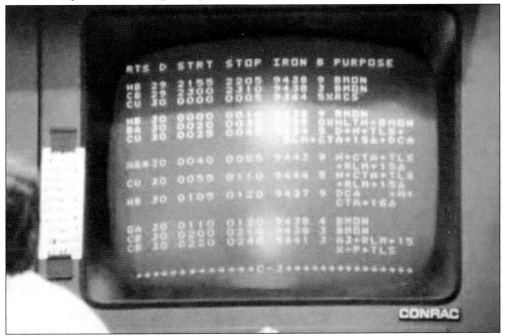

This screen shows how the AFSCF categorized its RTS contacts via the different antennae ("A-side" or "B-side") and inter-range operation number (IRON). By using an IRON, various organizations could refer to an individual satellite without betraying its launch date, booster, owner, or unique system designator. (USAF.)

This pamphlet from the 1985 Visitors' Day at Sunnyvale AFS provides a peek into the inner workings of the facility without revealing much about its past operations. The day was filled with tours, educational films, and trinkets from contractors like Ford Aerospace. (Author.)

The pamphlet map of Sunnyvale AFS shows the site layout that remained constant until the base closure in 2010. Buildings 1001 and 1003, along with the Sun East/West terminal building, were the only areas open for family tours. (Author.)

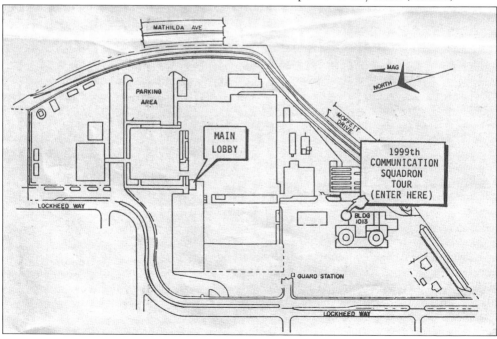

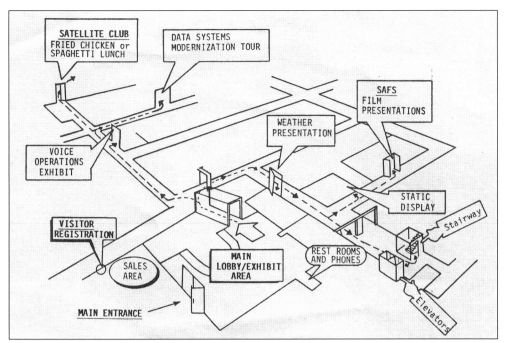

This side view gives the briefest indication of the layout of Building 1001. Family tours are great for community relations, but place a burden on facility workers who need to avoid any security incidents while 24/7 operations are ongoing. (Author.)

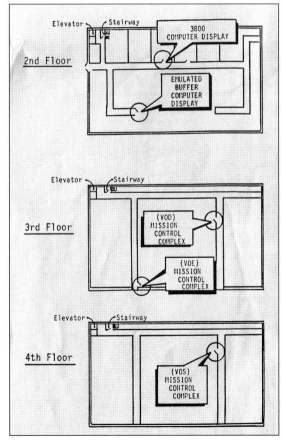

A less-detailed map of Building 1003 shows three floors and the directions to individual organizational sections for vehicle operations (VO). The designations were not described in detail within the pamphlet or inside the work centers. However, decades later, declassified materials would match the VO structure to its NRO alphabet programs and Operating Division Four (OD-4) successors. (Author.)

In the 1970s, the government mandated all military and civil satellites would use the space shuttle as their primary lift. One program on the shuttle's mid-1980s manifest was the Global Positioning System (GPS). After *Challenger*, the Department of Defense moved most of its satellites back to expendable rockets and eliminated its requirement for a military lift capability. The constellation of GPS satellites finally achieved completion in 1995 via Delta II expendable rockets. (USAF.)

A picture from STS-51-A signed by astronaut Joe Allen thanks Melvin Schuetz for assistance on satellite operations procedures. Schuetz worked at the Blue Cube, gaining valuable experience on satellite control procedures that transferred to his time as a senior satellite controller for Hughes on the Galaxy and Leased Satellite communications satellite programs. (Melvin Schuetz.)

Military dependents Joe and Erin Page (left and center) and Michelle Murphy, the smallest members of the Sunnyvale AFS team, are seen outside their base quarters. On January 19, 1983, a ribbon-cutting ceremony was held at Moffett Field as the first Air Force families occupied new military housing in Orion Park. Base housing intermixed Air Force families with their Navy counterparts, who supported Moffett's P-3 Orion submarine hunters. (Author.)

This c. 1987 view of NAS Moffett Field's Hangar One shows its distinctive black-and-white paint scheme. The airfield was built as an airship base, with the iconic Hangar One (1,133 feet long, 308 feet wide, and 198 feet high) as home to the Navy airship USS *Macon*. The dirigible hangar had doors at both ends of the structure, and the internal floor covered eight acres. (Author.)

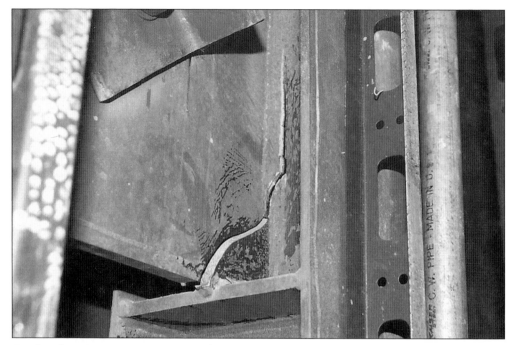

On October 17, 1989, at approximately 5:00 p.m., a 7.1-magnitude earthquake struck the Bay Area, with its epicenter near Loma Prieta Peak in the Santa Cruz Mountains. The earthquake caused major damage to parts of Building 1003, as shown here with a crack in the metal girder. (Onizuka Alumni.)

The 7.1 earthquake tossed offices around Onizuka Air Force Base. While initial damage was thought to be minor, extensive investigation revealed more than expected. Nearly $15 million was spent to provide seismic upgrades to Building 1003. (Onizuka Alumni.)

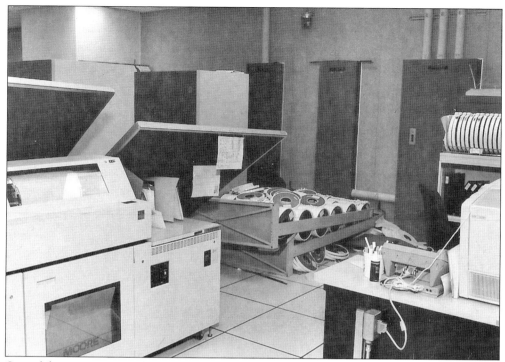

One of the extensive tape libraries inside Building 1003 was shaken up by the quake. The Loma Prieta earthquake reminded Air Force officials of the precarious placement of their primary satellite command and control node, leading to the movement of many operations to Falcon AFS, Colorado. (Onizuka Alumni.)

The Galileo space probe is released from the space shuttle *Atlantis* on October 18, 1989. Due to the diligence and fortitude of base employees, the base was able to provide launch support to the shuttle mission within 24 hours after the earthquake. (NASA.)

The Nation's Eyes and Ears in Space

Eight years after the declassification of the NRO, members of OD-4 created the *Millennium* yearbook to celebrate the accomplishments and legacy of the organization. While cryptically written in places, the book provided a nice pat on the back to the workers of Onizuka AFS. (Steve Olson.)

L to R: Chris Warrick, Joe Reft, Eric Dale and Steve Bose at the OD-4/DX 10th Anniversary Golf Celebration.

Celeste Ford and Joe Reft at Col Henley's going away golf tournament.

Members of OD-4 make "continuous improvement" a way of life. Even during their recreational hours, many unit members can be found at local golf courses performing "trajectory analysis", usually with a wide variety of results. OD-4 will use any excuse (launches, going away's, anniversaries) to swing the clubs.

OD-Fore

OD-4 Hits the Links

32 • Golf

In a display of levity within the OD-4 *Millennium* yearbook, the pages highlighting the love of golf shared by many workers at the site are humorously named "OD-Fore." (Steve Olson.)

Commanders of OD-4 are presented within the pages of the *Millennium* yearbook. The amount of experience obtained while working for the NRO is immeasurable, driving the organization to retain many of its best personnel. It is not uncommon to see young company-grade officers (lieutenants or captains) return to NRO staff positions as lieutenant colonels or colonels. Even after the closure of Onizuka, many faces within these pages could be found at other NRO locations. (Steve Olson.)

Tom Ludwig (deceased)
Commander of OD-4 from
1981-1984

Wesley West
Commander of OD-4 from
1984-1986

Jan Molvar
Commander of OD-4 from
1986-1990

Alex Alexander
Commander of OD-4 from
1990-1993

Kris Henley
Commander of OD-4 from
1993-1998

Jim Bailey
Commander of OD-4 from
1998-

OD-4 Supports
Habitat for Humanity

OD-4 Volunteers descended on Habitat for Humanity in Palo Alto and Redwood City to give something back to their communities. The volunteers work side-by-side with some of the future homeowners as they put in their sweat equity and learn the pride of home ownership. The critical componet here is the donation of time; no skills or tools required.

THE OD-4 CREW
in windows: Tim Rade, Tim Treglown. *back row:* Dave Madden, Dennis Farlow. *middle row:* Michael Fisher, Michael Walker, Kerri Mellor, Scott Larrimore, George Curcija, George Callow. *front row:* Linda Bellantoni, Renata Gross, Chris Povak, Ron Einhorn, Angie Ivey.

I LIKE MY REGULAR DAY JOB BETTER.
Linda Bellantoni helping out.

Workers from OD-4 support their local community through volunteer time with Habitat for Humanity. While military members rotate assignment locations many times throughout their careers, many civilian and contractor workers at Onizuka lived in the Bay Area permanently. Betterment of the community through these efforts helped solidify goodwill between the base and the local communities. (Steve Olson.)

Satellite schedulers at Onizuka stand at a table and edit the satellite tracking schedule with pencils and colored tape while coordinating with MCCs and RTSes. There were 10 people standing at a table measuring 40 feet working the schedule in real time or planning updates up to a week in advance. The paper scheduling method was replaced with the computer-based Automated Scheduling Tool for Range Operations in the 1990s. (USAF.)

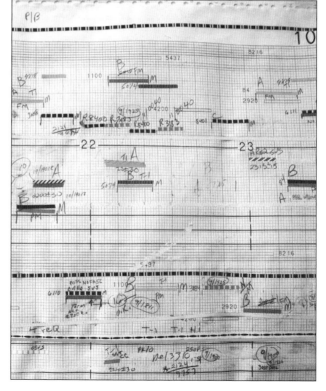

In this detailed view of the paper schedule built during operations in the early 1990s, large characters denote the A- or B-side at a specific RTS, while four-digit numbers throughout represent IRONs for a specific satellite. Adhesive tape with solid colors or hatch marks are placed throughout the schedule. Each square on the graph paper represents one minute of contact time. (USAF.)

Workers inside the Onizuka complex monitor their systems. These control rooms provided prelaunch, launch, early orbit, anomaly resolution, and operational Telemetry, Tracking, and Control support for specific missions. Note the logos of early space shuttle missions on the wall. (Onizuka Alumni.)

Maj. Thomas L. Hickerson addresses a formation of airmen after assuming command of the 1999th Communications Squadron (CS) on July 11, 1985. The 1999th CS mission was absorbed into the 21st Space Operations Squadron (SOPS) by 1992 along with the responsibilities of the 2nd Satellite Tracking Group. The advanced technology used at the AFSCF necessitated a dedicated cadre of communication experts. (National Archives.)

One of Onizuka's CDC-3800 computers resides in the National Air and Space Museum's Udvar-Hazy Annex in Chantilly, Virginia. The system used 48-bit word length with logic consisting of discrete transistors. The 3800 held 128 kilobytes of memory and cost nearly $1.9 million. In 1972, Building 1003 held seven CDC-3800s. (National Air and Space Museum.)

Banks of batteries support Onizuka's uninterruptible power supply system, necessary when primary power to the facility was disrupted. The batteries would carry the electrical load until Building 1004's electrical generators came online to provide power. (Onizuka Alumni.)

A late 1990s coversheet provides information within the Byeman Control System. Seen as necessary during the early days of the reconnaissance program, many of the programs within the control system were consolidated into a single large security compartment to better support soldiers. The success during Desert Storm found military commanders recognizing the value of NRO information on the battlefield. One declassification effort saw the NRO being openly acknowledged in 1992. (NRO.)

Workers pose in front of the 51-foot Data Link Terminal antenna during its dismantling in April 2009. The dish required two counterweights to stabilize its motion during operations. Disassembly of the dish required over 1,200 man-hours and use of a 210-ton construction crane. During disassembly, workers often had to stop work to allow the Valley Transportation Authority's light-rail train to pass nearby. (Onizuka Alumni.)

Workers inside the configuration facility provide IRON interface instructions for specific antennas. When the system was installed in the 1970s, it was considered state of the art, with eight-inch, 56-kilobyte floppy disks transferring information to the RTS locations. By the 1990s, sustainment of the outdated floppy disks and disk drives was untenable, leading the Space and Missile Systems Center to replace outdated equipment with network interface terminals. (Onizuka Alumni.)

Outside of Building 1002, a sign (above) displays the motto of 21st SOPS ("Gateway to the Stars"). Next to the sign is a wall with a color tile mosaic (below) representing the various missions supported by Onizuka AFS over the decades. From left to right are an IUS booster, a GPS satellite, a Defense Meteorological Satellite Program weather satellite (top), and the space shuttle with an astronaut spacewalking. Noticeably absent are representative images of the reconnaissance satellites supported since the station's inception in the late 1950s. (Both, author.)

The S curve along Innovation Way was closed to all traffic in the late 1990s due to increased force protection measures to maintain a stand-off distance from operational areas. The artificial grass was added in 2004 to enhance the perimeter's appearance. (Onizuka Alumni.)

The heritage display of Ellison Onizuka's life resided in the lobby of Building 1001. These items represented the largest display of Onizuka's personal effects outside of his immediate family. Photographs from his life, career, and previous spaceflight aboard STS-51-C filled the cases. (Onizuka Alumni.)

An oil painting of Ellison Onizuka was completed by Manuel Jaramillo Rodriguez on August 28, 1986, soon after Sunnyvale AFS was rededicated. Onizuka wears the silver oak leaves of a lieutenant colonel because when the painting was completed, his posthumous promotion to full colonel (on November 1986) had not yet been signed into law. (USAF.)

In May 2010, members of the 21st SOPS family pose for photographs during the final days of the facility, both at the main entrance (above) and the southern parking lot (below). As home to thousands of contractors, military, and government civilians over the decades, the location holds a special place in the hearts of those who worked there. The photograph above shows many of the iconic locations of the base, including the Blue Cube, Sun East and Sun West satellite dishes, and power plant. (Both, Onizuka Alumni.)

The people here, photographed during the NRO's 50th-anniversary gala in 2011, were all assigned to NRO programs in Sunnyvale. The personnel wearing the distinctive white ribbons were veterans of the Gambit and/or Hexagon programs at various times. (Melvin Schuetz.)

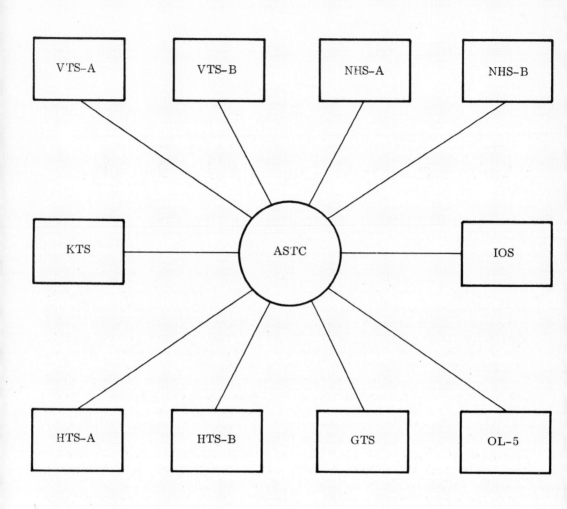

During the ASTC development, the entire satellite control network consisted of seven tracking stations. The A and B designators denote separate antennae at a particular ground station, highlighting independent capability to track and relay data from separate satellites. The ASTC design, with its MCCs and Advanced Data System "Bird Buffer" data queuing system, helped mature technology for the present-day AFSCN. (Tom Carr.)

Four

GATEWAY TO THE STARS

No engineers present during the development of the WS-117L Subsystem H could have imagined how widespread, both figuratively and geographically, their initial design would become. Starting as a control node with three remote ground stations, the network would blossom into three control nodes (two after Onizuka's closure), 10 "true" RTSes, and other interface sites. Building on the success of the initial Department of Defense and NASA missions using the control and tracking stations, the network grew into its own program of record, the AFSCN.

The AFSCN's primary mission is to provide communications links to military, civil, and national satellites. Through these connections, satellite operators provide force-multiplying effects for the military with continuous global coverage, low vulnerability, and autonomous operations. Satellites serviced through the AFSCN's common user element provide in-theater secure communications, weather, and navigational data for ground, air, and sea operations.

Changes to the AFSCN began in 1979, when the secretary of defense authorized development of the Consolidated Space Operations Center (CSOC) in Colorado Springs, originally consisting of two parts—the Satellite Operations Complex and the Shuttle Operations and Planning Center (SOPC). The loss of *Challenger* saw the SOPC cancelled, leaving CSOC as a satellite control location mirroring the capability of Onizuka. Air Force Space Command's push to operational space assets saw become the control location for most Department of Defense satellites; the Air Force Satellite Control Facility (in name only) was inactivated in 1987, returning the facility to its research and development roots with a new name mirroring an old one: the Consolidated Space Test Center (CSTC).

Tracking stations supporting the AFSCN's mission are found in areas of extreme juxtaposition, from the negative temperatures at the extreme latitudes of Thule Air Base, Greenland, to the tropical lushness of Diego Garcia, Guam, and Hawaii. No matter the hardship of working at these remote locations, members of the AFSCN family recognize their support to such an important mission for the US Air Force and the nation.

Bob Siptrott, document control operator at the Kodiak Tracking Station, poses with a rifle in front of a bearskin around 1968. Kodiak operators had to transmit data back to Sunnyvale around 1/32 speed, requiring 32 minutes to send one minute of telemetry. Occasionally, civilian Sunnyvale phone operators would hear random beeps and squeaks and disconnect the in-use lines, requiring the RTS to retransmit the data. (Kadiak.org.)

Operating Location 3, on Annette Island, Alaska, was a short-lived RTS in the network, obtaining full operational capability on May 25, 1962. While Annette Island received the Multi-Satellite Augmentation ("Augie") Program upgrade in 1962, it was deemed excess and was deactivated in the first half of 1963. (USAF.)

Starting out as Operating Location 9, the Indian Ocean Station RTS on Mahe, Seychelles, logged its first contact in October 1963. Worldwide constellations, such as the Vela Hotel nuclear detection satellites, required continuous contact with ground stations. The site supported continuous operations unabated until 1996, though the years were not without drama. (USAF.)

Prior to gaining its independence in 1976, the island operated as a British colony. Internal political strife throughout the late 1970s and 1980s within the Seychelles government, including multiple coups, kept American personnel constantly on edge. American negotiators balked at proposed lease hikes, such as an exorbitant $10-million-per-year fee (in 1980 dollars). The agreement for the site ended in 1993, with the station being decommissioned in 1996. (USAF.)

The 60-foot dish at the Camp Parks Communications Annex is under construction in 1960. The facility provided launch and early orbit functions and calibration activities for new satellites. Testing facilities were previously operated by MIT Lincoln Laboratory before becoming part of the network in 1970. The site converted to contractor operation in 1978. (USAF.)

On July 1, 1970, the AFSCF assumed responsibility for the Camp Parks Radiometric Test Facility in Dublin, California. Prior to becoming a part of the AFSCF, Camp Parks provided housing for officers and enlisted personnel working in Sunnyvale. An early wideband communications system was tested at Camp Parks; it would provide connectivity up to 1.5 megabits of data per second to terminals at the RTSes. (USAF.)

The Colorado Tracking Station was opened in 1988 on Falcon AFS in Colorado Springs. The RTS was located near the CSOC, giving the complex a direct connection to the network. The site's proximity gave Falcon AFS enhanced capability beyond Onizuka's, since the Sunnyvale site did not have RTS capability onsite. (USAF.)

Demolition crews dismantle an antenna building on September 19, 2014, at the Colorado Tracking Station on Schriever Air Force Base. When it was constructed, the site was equipped with Automated Remote Tracking Station technology, creating a conduit from satellites to operators. Prior to this upgrade, tracking stations were commanded onsite, which meant that satellite operators had to relay commands to antenna operators, who then transmitted them to the satellites. (USAF/Dennis Rogers.)

An agreement with the United Kingdom in the 1990s allowed placement of a tracking station on Diego Garcia in the Chagos Archipelago, British Indian Ocean Territory, to make up the expected coverage loss of the station in the Seychelles. It was accepted into the network in the early 1990s. The site also included a GPS tracking station, part of a network independent of the AFSCN. (USAF.)

M.Sgt. David Perl poses for a photograph outside of the ops center on the atoll of Diego Garcia. The AFSCN's shift from research and development to an operational mindset placed the RTSes under the watchful eye of Air Force Space Command's Office of the Inspector General to ensure compliance with Air Force standards and regulations. (David Perl.)

Operating Location 10 at Guam's Northwest Field opened on September 17, 1965. Originally configured as a mobile facility, the site consisted of seven vans for administrative support, seven for power, one for maintenance, and nine for technical operations. The mobile tracking station was officially deactivated in October 1969 in favor of a permanent facility that entered the AFSCF inventory seven months earlier. (USAF.)

Sr. Amn. Alvin Alvarez (left) and A1c. Robert Young check the perimeter at the Guam Tracking Station. The airmen provide security for the station 24 hours a day, reflecting a radical paradigm shift from the early days of the Air Force's space program. The white Kevlar-covered environmental dome behind them is rated to withstand 195-mile-per-hour typhoon winds and adapts its internal pressurization based on outdoor wind speed calculations. (USAF.)

Four airmen work on tracking consoles at Kaena Point, Hawaii. The tracking station at Kaena Point was a critical link in the nascent satellite control network. During a commanding re-entry, the site would track the speed and projected final location for the film capsules and relay information to the "Star Catchers" of the 6594th Recovery Control Group at Hickam Air Force Base. (USAF.)

Located on the western tip of O'ahu at an elevation of over 1,500 feet, Kaena Point Tracking Station was one of the earliest sites in the WS-117L network, constructed in 1959. In ancient times, the site was known as "the leaping place of souls" (*leina a ka'uhane*), where the spirits of the recently dead could be reunited with their ancestors. (USAF.)

Selected as the northeast control site in the WS-117L network, the New Hampshire Tracking Station was accepted by the Air Force on April 1, 1960. The location was part of a World War II bombing range for aircraft from nearby Grenier Army Air Field, New Hampshire. In little over a year, the site received the Multi-Satellite Augmentation Program upgrade, allowing 24/7 support for up to five different types of satellites at the same time. (USAF.)

In October 1987, the New Hampshire Satellite Tracking Station was assigned under Air Force Space Command, reflecting the AFSCN's move from Air Force Systems Command's research and development role to an operational environment. The New Hampshire location boasts some of the best recreational lands held by the Air Force, with cabins, lakes, and over 25 camping sites. Unexploded ordnance from World War II is occasionally found within its 3,000 acres. (USAF.)

Royal Air Force Base Oakhanger began its association with the US Air Force as Operating Location AE of the AFSCF in 1978. The site provides support to US civil and national satellites. It also performs the unique function of supporting the United Kingdom's Ministry of Defence Skynet satellite communications constellation, which supports the three branches of the British armed forces. (USAF.)

Located near a village in the English countryside, RAF Oakhanger is the only AFSCN RTS primarily operated by non-American personnel; however, a small contingent of US Air Force personnel resides there. The site's location provides dual coverage for geosynchronous satellites near the Atlantic and Indian Oceans, supplementing the coverage of Diego Garcia and Thule. (USAF.)

During the early days of the reconnaissance satellite program, the placement of Operating Location 5, also known as POGO, was classified due to political sensitivities. Located at Thule Air Base, Greenland, POGO was named after its planned primary mission—supporting NASA's Polar Orbiting Geophysical Observatory. After an NBC newscast in 1964 and numerous mentions in contracting information and the base phonebook, the AFSCF declassified the location in 1969. (USAF.)

Technicians work at the Thule Tracking Station around 1981. Thule has the most antennae of any AFSCN RTS due to the sheer number of daily contacts and specialized mission support. The site is almost 700 miles above the Arctic Circle and just 1,000 miles south of the North Pole. The site is in constant darkness from November to February and constant sunlight from May to August. (USAF.)

As one of the original RTSes for the Discoverer program, Vandenberg Tracking Station became operational in 1959. Its call-sign (COOK) came from the base's original name of Cooke Air Force Base, which stemmed from its World War II–era Army heritage as Camp Cooke. In 1966, the site became a demonstration station along with Thule for the revolutionary SGLS data transmission system. (USAF.)

Vandenberg Tracking Station resides atop a hill overlooking the Pacific Ocean almost 950 feet above sea level. This altitude provides excellent coverage for the unit's secondary mission of providing launch infrastructure augmentation to Western Range missions. Before a satellite is launched from Vandenberg, signals transmitted toward the site's antenna dishes verifies AFSCN connectivity; if troubleshooting is required, it can be performed before launch. (USAF.)

Five

ENIGMATIC EMBLEMS

Emblems have been used to represent military units from the Roman Empire's *aquila* (eagle) symbol to the modern military's unit patches. Official Air Force insignia are designed by the US Army's Institute of Heraldry (TIOH) at Fort Belvoir, Virginia; after creation, an Air Force emblem's heraldic symbology remains on file at the Air Force Historical Research Agency at Maxwell Air Force Base, Alabama.

Many of the symbols used by units attached to the operations at Sunnyvale/Onizuka represent the units' off-world mission. Delta symbols represent space-related hardware; delta-v is the measure of the change in velocity used in spacecraft flight dynamics. Stars may represent pieces or parts of a unit's mission, such as RTS from the AFSCN, or numbers of spacecraft in a series. Elliptical shapes often represent orbital paths around the earth.

While the official NRO seal design is governed by TIOH regulations, many patches representing launches, satellites, and organizations have sprung up over the decades. Due to the organization's classified nature, few if any logos or emblems exist for the period from 1961 through 1992. In 1996, however, the NRO began announcing launches on a regular basis. Since then, images representing launches have appeared and stupefied "patchologists," amateur space history observers attempting to decode the symbology inside the emblems.

While a portion of this chapter's emblems are official, representing heraldic work from TIOH, the remainder came from printed materials obtained from former Onizuka personnel, such as an OD-4 yearbook and an unpublished Onizuka history. While their official status may be in doubt, the designs reveal a conundrum within the classified world: how to represent your organization without revealing details of its operations.

The logo of the 6594th Test Wing shows a bird of prey circling a globe amid a sea of stars. The motto *Inveniemus viam vel faciemus* translates to "We will find a way or we will make one." This phrase has been apocryphally attributed to Carthaginian general Hannibal Barca after being told by his generals it was impossible to cross the Alps by elephant. (AFHRA.)

The 2nd Satellite Tracking Group operated the AFSCN and commanded on-orbit satellites in support of Department of Defense, NATO, and NASA programs. The unit was organized under the 2nd Space Wing (at Falcon AFS, Colorado) and represented a shift from developmental/test programs under Air Force Systems Command to operational programs under Air Force Space Command. (AFHRA.)

The 6593rd Test Squadron (Special) was organized on August 1, 1958, at Edwards Air Force Base, California. After developing procedures for the aerial recovery of capsules ejected from orbiting satellites, the unit moved to Hickam Air Force Base, Hawaii, in November 1959, and was placed under the 6594th Recovery Control Group. The unit logo shows an Air Force blue falcon in flight reaching its talons toward a red parachute, a subtle nod to its assigned mission. (AFHRA.)

"Catch a Falling Star," the motto on the 6594th Test Group, provided a veiled description of the unit's mission. The star represents a re-entering space vehicle "captured" by golden falcon wings on either side. The 6594th Test Group flew 10 different aircraft types over 28 years (1958–1986). The unit was inactivated soon after the last (failed) Hexagon launch on April 18, 1986; technology had progressed beyond the need for film capsule recovery. (AFHRA.)

The 5th Space Operations Squadron conducted command and control for a number of civil and military programs: NATO III/IV, Skynet, and Defense Satellite Communication System satellites, as well as launch and early orbit support for the space shuttle and the IUS booster. The unit's various missions were absorbed by the 21st SOPS in June 2000. (AFHRA.)

The mission of the 21st SOPS is to "protect, monitor, conduct and enable satellite and launch operations in support of national objectives." The unit was activated in October 1991 and absorbed the responsibilities of the 750th Space Group and 5th SOPS after the 1995 BRAC determination recommending realignment of Onizuka. After closure of the site, the unit moved to Vandenberg Air Force Base. (AFHRA.)

The 750th Support Squadron emblem and motto state the importance of Onizuka to the satellite command and control network. The hemispherical environmental dome and orbiting delta emblem aptly represent the functions performed by AFSCN RTSes around the world. (National Archives.)

The NRO organizational seal evolved out of a design for the Office of Space Systems, Office of the Secretary of the Air Force, a cover organization. When the NRO presented briefings inside classified locations, the briefers replaced the name of the cover organization with the true name, National Reconnaissance Office. The NRO adopted the emblem as its official seal in 1994. (NRO.)

Operating Division Four (OD-4) was an NRO organization at Sunnyvale. Very little has been declassified about OD-4's missions or operations. For a time, Sunnyvale AFS also housed an Operating Division One (OD-1) before consolidation with OD-4 in 1994. The emblem contains typical heraldic elements representing space (stars, delta, and orbital paths) and one cube-like structure at center left representing Building 1003. (Author.)

OD-4/DC has the longest historical legacy at Sunnyvale. Established after a successful reconnaissance launch in July 1963, it ran numerous "90-day wonder" programs (satellites with 90-day warranties). As lifespans for reconnaissance satellites increased, DOZC, and later VOC, provided near-real-time support. The patch's design shows polar orbits, common for many reconnaissance missions, as well as the tongue-in-cheek motto "We Bring You the World." (Author.)

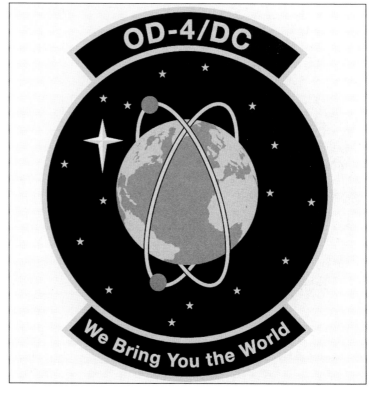

OD-4/DX was originally named VOX and was established in the early 1980s before being absorbed into OD-4. The owl motif, motto of "We Own the Night," and geometric features around the globe have given amateur patchologists plenty of research fodder. DX recognized its 10-year anniversary in 1999, inviting past and present team members back to celebrate the group's accomplishments. (Author.)

The patch for OD-4/DH shows interesting orbital geometry along with a pair of dice displaying the numbers seven and eleven. OD-4/DH began as VOH at the AFSCF in 1976. It was originally located on the fourth floor of Building 1003 until moving to the second-floor extensions in 1986. DH personnel pioneered the development of the Data System Modernization Program, which became the standard system for satellite command and control. (Author.)

This logo emphatically identifies the Air Force's Sunnyvale facilities, including the Blue Cube, as the "Birthplace of Space Reconnaissance." While detractors have dismissed the statement as hyperbole, the creation of the Agena upper stages by Lockheed and the system's criticality to the Corona, Gambit, and Earpop programs have cemented Sunnyvale in the annals of space history. (Author.)

Six

PASSING THE TORCH
2010–2016

At the start of the 21st century, the stage was set for Onizuka's closure. Increasing mission requirements for both NRO and Department of Defense satellites kept outpacing what Sunnyvale personnel could handle with its aging infrastructure. While the 1995 BRAC realignments slashed Onizuka's workforce by nearly 3,000 jobs, the 2005 BRAC decision recommended the dreaded "C"—closure. The NRO continued its responsibilities, with control of its satellites spread throughout mission ground stations, while Schriever Air Force Base, Colorado, took on the responsibility as the AFSCN primary node, with Vandenberg Air Force Base as a backup.

On March 7, 2007, Dr. Donald M. Kerr, director of the NRO, spoke about the site's legacy during OD-4's closing ceremony:

> Over the last 46 years, the site quietly, but effectively, discharged its responsibilities, and became a major asset of the intelligence community, supporting senior civilian decision makers and our military forces in both war and peace. It forged a close and unique relationship between the Air Force and the intelligence community that set standards for responsiveness and operational excellence. It is no coincidence that a number of leaders, past and present, in the intelligence and reconnaissance communities spent time at this site.
>
> Today, we take the opportunity to recognize the contributions of this unique facility and those who worked there. In fact, we can say without the slightest qualification that the nation's preeminence in spaceborne reconnaissance owes much to the men and women of the Blue Cube.

The City of Sunnyvale considered various alternatives for the reuse of the property: homeless housing, an auto dealer mall, and offices for Veterans Affairs. It was a local community college, however, that won the land grab. In 2012, the Foothill-DeAnza Community College District obtained 9.15 acres of the 18.9-acre Onizuka property free of charge from the US Department of Education. The Lionakis architectural firm designed the Foothill College Sunnyvale Center, a multistory, 50,000-square-foot state-of-the-art instructional facility incorporating the color blue and a cube-like design. The facility houses training for emergency medical technicians, child development, and geospatial technologies. The legacy of the Blue Cube still resonates through the campus.

Col. Wayne Monteith, the 50th Space Wing commander, delivers a speech under Sun East and Sun West during the closure ceremony of Onizuka Air Force Station on July 28, 2010. Primary responsibilities for the AFSCN continued under the 50th Space Wing's 22nd SOPS at Schriever Air Force Base, Colorado. (USAF/Sr. Amn. Bryan Boyette.)

Lorna Onizuka speaks at the closure ceremony of Onizuka AFS on July 28, 2010. After the *Challenger* accident in 1986, she remained active in space program activities, becoming a consultant to the Japanese Aerospace Exploration Agency and a founding director for Challenger Center, encouraging junior high school students to cultivate skills needed for future success, such as problem solving, critical thinking, communication, and teamwork. (USAF/ Sr. Amn. Bryan Boyette.)

The Travis Air Force Base Honor Guard (in dark uniforms) sounded retreat and retired the American flag over Onizuka AFS for the final time on September 15, 2011. Rendering honors during the retreat ceremony are, from left to right, 2nd Lt. Nick Krieger (behind flag), Technical Sgt. Russ Boring, 50th Space Wing Network Operations Group commander Col. Michael Finn, Detachment 4 site commander Lt. Col. Andy Wulfestieg, and former Detachment 4 chief Rose Beamer-Jansson. (USAF.)

On July 30, 2010, Vandenberg Air Force Base personnel held a ribbon-cutting ceremony at the Ellison Onizuka Satellite Operations Facility, the new home of the 21st Space Operations Squadron. The presiding official for the ceremony was the 14th Air Force commander, Lt. Gen. Larry James. As part of the manned spaceflight engineer program, then-Captain James would have been intimately familiar with the AFSCF's operations of the early 1980s before the program's cancellation in 1988. (USAF/Sr. Amn. Bryan Boyette.)

The oil painting of Col. Ellison Onizuka by Manuel Jaramillo Rodriguez sits outside during the ribbon-cutting ceremony at the new facility bearing his name. As the AFSCN backup node, site personnel schedule, allocate, and configure common user resources while monitoring the status of worldwide AFSCN resources. Operations supported include missile warning systems, meteorological data, NASA launch and re-entry missions, and intelligence-related programs. (USAF/Sr. Amn. Bryan Boyette.)

Lorna Onizuka speaks at the ribbon-cutting ceremony at Vandenberg Air Force Base on July 30, 2010. She stated, "I have slowly absorbed what you all do. . . . Things I admit I did not know. I am tremendously impressed. I sincerely salute all of you for the work that is done here. You do invaluable work to ensure the security of our nation—and we are grateful." (USAF/Sr. Amn. Bryan Boyette.)

Lt. Col. Andy Wulfestieg, commander of 21st SOPS, gives a tour of the Ellison Onizuka Satellite Operations Facility during the unit's 21st birthday celebration on October 5, 2012. Wulfestieg commented, "This party is meant to celebrate our past accomplishments, while highlighting our future endeavors through the growth of technology and those relationships that are built with other government agencies." (USAF/Levi Riendeau.)

The centerpiece of the heritage display inside the Ellison Onizuka Satellite Operations Facility at Vandenberg Air Force Base is the oil painting by Manuel Jaramillo Rodriguez, part of the US Air Force Art Collection. All of Colonel Onizuka's personal items from Sunnyvale were brought down to Vandenberg for display inside the new facility. (USAF/Michael Peterson.)

A photograph of El Onizuka in his Air Force ROTC uniform is displayed next to his University of Colorado student identification card and ROTC ribbons. Before the advent of desktop calculators, the slide rule was the most commonly used calculation tool in science and engineering disciplines. This one likely served Onizuka well in his undergraduate and graduate aerospace engineering classes. (USAF/Michael Peterson.)

The display case contains a commemorative plaque with two patches (STS-51-L and subdued AFSCF logo) that were flown as part of *Challenger*'s flight kit and retrieved after the accident. Colonel Onizuka's medals are presented alongside the photograph of the STS-51-C crew outside of the Satellite Club (Consolidated Open Mess) at Sunnyvale AFS in late 1984. (USAF/Michael Peterson.)

Inside the Onizuka Satellite Operations Facility, a photograph of Colonel Onizuka with an infectious smile is matted next to a launch photograph and a crew patch for STS-51-L. The picture was dedicated to Onizuka Air Force Base and signed by his widow, Lorna Onizuka. (USAF/Michael Peterson.)

The standard US Air Force restricted area perimeter warning sign remains attached to the metal fencing on Lockheed/Innovation Way during the Blue Cube's demolition in 2014. The short distances from the perimeter to the AFSCF buildings underscored force protection concerns in the 1980s, necessitating creation of a backup facility at Falcon AFS (later Schriever Air Force Base) in Colorado Springs. Ironically, the Colorado backup facility would later supplant most of Onizuka's capabilities. (Ian Abbott.)

Slight discoloration in the bottom concrete panels indicates this portion of the building was connected to Building 10031, the three-story extension constructed in 1984 to provide two stories of parking space and a single story for mission operations. Displaying a rare symbol of levity, a graffiti artist calling himself John proclaims his "luv" for Jane on the north side of Building 1003 (lower center, near doorway). (Ian Abbott.)

In a reversal of its construction, Building 1003's concrete panels are removed before demolition of the building's guts. A sign for construction company C.W. Driver hangs from the top. No notification was given to the public by the company before final demolition began on April 16, 2014. (Randy Rhody.)

A crane sits in front of the remnant of Building 1003, the famed Blue Cube. In a story about the site's demolition, *San Jose Mercury News* reporter Becky Bach discovered that the destruction began during hours of darkness—late evening or early morning—in a "fitting end" to the secretive Cold War installation. (Randy Rhody.)

A close-up view inside Building 1003 shows wires used to hold up false ceilings and air-conditioning ducts. The amount of computer processing power (five CDC-3800 computers) originally installed in the building likely paled in comparison to the computing power used during its final years. Many computer upgrades occurred over the decades, such as the Advanced Data System and Data Modernization Program, requiring more resources to cool equipment. (Randy Rhody.)

In this side view during Building 1003's demolition, the distinct plenum space between floors two and three can be seen. While the external connections between the Blue Cube and the RTS locations were important, the intra-facility data lines were just as critical, providing external telemetry as input to the banks of processing computers juggling the needs of numerous MCCs. (Randy Rhody.)

Security doors removed from the building sit in a pile awaiting final disposition. Since many areas inside the buildings at Onizuka required security clearances, the high-security doors are not surprising. Even within the high-security facility, areas were compartmentalized from other programs being run nearby. (Ian Abbott.)

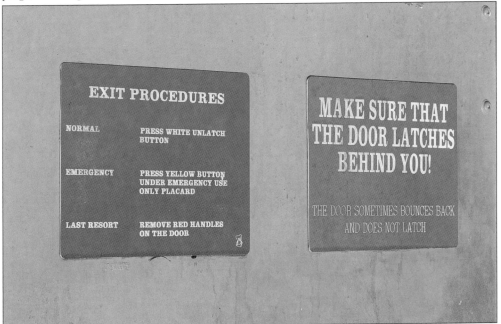

EXIT PROCEDURES

NORMAL	PRESS WHITE UNLATCH BUTTON
EMERGENCY	PRESS YELLOW BUTTON UNDER EMERGENCY USE ONLY PLACARD
LAST RESORT	REMOVE RED HANDLES ON THE DOOR

MAKE SURE THAT THE DOOR LATCHES BEHIND YOU!

THE DOOR SOMETIMES BOUNCES BACK AND DOES NOT LATCH

Information signs on the discarded security doors seem to echo common-sense information to a normal user. In a panic during an emergency, however, safety instructions will be forgotten, ignored, or performed incorrectly, hence the simplified directives. (Ian Abbott.)

These concrete panels are rarer than moon rocks and prized by space memorabilia aficionados. No advance notice was given by contractor C.W. Driver before the demolition of Building 1003. However, one enterprising forward-thinker obtained small portions of the concrete panels off the Blue Cube. (Jerry Race.)

A few of the large concrete panels were sliced into miniature "Baby Blue Cubes." While the blocks are nothing more than blue paint, cement, and aggregate, they will forever memorialize the missions and people that occupied the mysterious building off of Mathilda Avenue. (Jerry Race.)

The Foothill College Sunnyvale Center sits directly atop the former location of Building 1003, better known as the Blue Cube. The location offers career training for emergency medical technicians, child development specialists, and geospatial technology. The site contains many features that harken back to the legacy of Onizuka AFS. (Author.)

The front of the Foothill College Sunnyvale Center sits along Innovation Way off of Mathilda Avenue. Foothill College was given nine acres of real estate by the federal government. The external features of the building hint at the land's previous resident, with a cube-like appearance displaying a distinctive blue color. (Author.)

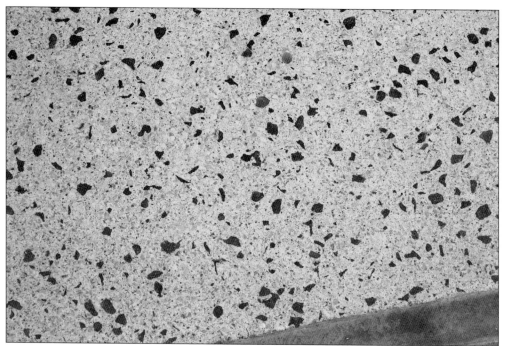

The legacy of the Blue Cube remains inside the bones of Foothill College Sunnyvale Center, both figuratively and literally. Small blue reflective minerals, reminiscent of the Blue Cube's facade, are interspersed in the concrete walkway leading to the main entrance. (Author.)

The Foothill College Sunnyvale Center mascot is the owl, as seen on this t-shirt. The campus coffee shop offers a small variety of items with the owl mascot and is located across from the campus's heritage room, which houses items from Onizuka's past. (Author.)

The owl mural inside the staircase of the Foothill College Sunnyvale Center uses imagery reminiscent of the patch from the NRO's OD-4/DX section. DXers, known as the "Night Owls," supported military activities around the world, including Kosovo, Desert Storm, and air strikes in Afghanistan. (Author.)

The names of former Blue Cube employees surround the owl mural inside the Foothill College Sunnyvale Center. Most of the names include the years of employment at the base, while some include humorous in-jokes, such as "Enter the Dragon" under the name Bruce Lee at lower right. (Author.)

Another icon is lost. In 1997, personnel at NASA Ames Research Center discovered polychlorinated biphenyls (PCBs) in their storm drain settling basin. The source of the PCBs was determined to be the materials used to create Hangar One's external sidings. As a result, Hangar One was closed to human use. Tech giant Google has pledged to reskin Hangar One and build an educational facility inside the historic landmark by 2025. (Author.)

Time has taken its toll on the *Columbia* memorial garden. As of 2018, the wooden stumps remain in place, though all the STS-107 crew member nameplates are missing. It is unknown if a similar memorial was recreated at the Ellison Onizuka Satellite Operations Facility at Vandenberg Air Force Base. (Fred Taleghani.)

The control consoles for the Solar Saturn gas turbine generators await disposal. As originally envisioned, the gas turbines were designed to operate as a total energy system for the STC and provided all the electrical and mechanical power required to support Buildings 1001 and 1003. Since Building 1004 was critical to ensure continued support of the base's satellite programs, all power generation systems were designed with redundancies. (Fred Taleghani.)

These concrete pads were oversized pedestals for the iconic Sun East and Sun West satellite antenna dishes. As of 2018, the raised platforms were still present, succumbing to the environment. In the distance, a lone satellite dish remains, mission and customer unknown. (Author.)

During the distribution of former Onizuka property, the Department of Veterans Affairs (VA) obtained 4.4 acres, including Building 1002. The 51,000-square-foot structure required seismic retrofitting to meet present earthquake standards in preparation for a renovation and lobby addition. (Author.)

This sign in front of Building 1002 is one of the few external indicators that remain at the VA complex. (Author.)

What was done by one is now performed by many. With the demand for national reconnaissance products increasing in the past few decades, the NRO has expanded Onizuka's missions to multiple ground stations. A handful of ex-Onizuka employees have taken their knowledge and experience to these locations. These aerospace data facilities (ADFs) are multi-mission ground stations responsible for supporting worldwide defense operations and multi-agency collection, analysis, reporting, and dissemination of intelligence information. They provide data to defense, intelligence, and civil agencies supporting the US government and its allies. Clockwise from top left are logos for ADF-Colorado, ADF-East, ADF-Southwest, and the National Reconnaissance Operations Center. (NRO.)

Every generation has the obligation to free men's minds
for a look at new worlds . . . to look out from a higher
plateau than the last generation.

Ellison S. Onizuka

This document contains sensitive
electronics. For best performance,
do not bend, perforate or expose
to extreme temperatures.

A. RESTRICTIONS ON IMPORTATION OF GOODS AND SERVICES For
information, write the Department of the Treasury, Office of Foreign Assets
Control, Treasury Annex, 1500 Pennsylvania Avenue N.W., Washington, D.C.
20220, or consult http://www.treas.gov/ofac

B. CUSTOMS & BORDER PROTECTION Contact Customs & Border Protection
(CBP), for a copy of "Know Before You Go" and "Pets, Wildlife - Licensing and
Health Requirements," at http://www.cbp.gov/xp/cgov/travel/

C. AGRICULTURE For a copy of "Travelers' Tips On Bringing Food, Plant,
and Animal Products into the United States,"contact the U.S. Department of
Agriculture, http://www.aphis.usda.gov/travel See also U.S. Fish and Wildlife
Service, http://www.le.fws.gov/TipsforTravelers.htm for other important
"Information for International Travelers."

D. U.S. TAXES All U.S. citizens working and residing abroad are required
to file and report on their worldwide income. Consult IRS Publication 54,
"Tax Guide for U.S. Citizens and Resident Aliens Abroad," available at
http://irs.gov/publications/p54/index.html

E. SOCIAL SECURITY Write to the Social Security Administration, Office of
International Operations, P.O. Box 17769, Baltimore, MD 21235; or consult
http://www.ssa.gov/international about receiving payments while outside the
U.S., or contact the nearest Social Security office in the United States or at a
U.S. embassy or consulate abroad.

EXACT WEBSITE ADDRESSES SUBJECT TO CHANGE.

Every US passport contains a quote from Ellison Onizuka's 1980 commencement address to Konawaena High School: "Every generation has the obligation to free men's minds for a look at new worlds . . . to look out from a higher plateau than the last generation. Your vision is not limited by what your eye can see, but what your mind can imagine. If you accept these past accomplishments as commonplace, then think of the new horizons that you can explore. . . . Make your life count, and the world will be a better place because you tried." (Author.)

BIBLIOGRAPHY

Arnold, David. *Spying from Space: Constructing America's Satellite Command and Control Systems.* College Station, TX: Texas A&M University Press, 2008.

Berger, Carl. *History of the Manned Orbiting Laboratory Program (MOL).* Washington, DC: MOL Program Office, 1970.

Bradburn, David, John Copley, and Raymond Potts. *The SIGINT Satellite Story.* Center for the Study of National Reconnaissance. Chantilly, VA: National Reconnaissance Office, 1994.

Butterworth, R. *The U.S. Signal Intelligence Satellite Enterprise: A History.* Center for the Study of National Reconnaissance. Chantilly, VA: National Reconnaissance Office, 2005.

Harland, David M. *The Story of the Space Shuttle.* London, UK: Springer-Verlag, 2014.

Jernigan, Roger. *Air Force Satellite Control Facility: Historical Brief and Chronology, 1954–Present.* NRO WS-117L Records Collection. Sunnyvale, CA: Air Force Satellite Control Facility History Office, 1983.

Malinowski, Tonya. "Mission Accomplished." ESPN E:60, June 29, 2018. www.espn.com/espn/feature/story/_/id/23902766/nasa-astronaut-ellison-onizuka-soccer-ball-survived-challenger-explosion/.

National Reconnaissance Office. *Declassified Records Collection.* Information Access and Review Team, 1995–2017. www.nro.gov/Freedom-of-Information-Act-FOIA/Declassified-Records/.

Oder, Frederic, James Fitzpatrick, and Paul Worthman. *The Corona Story.* Center for the Study of National Reconnaissance. Chantilly, VA: National Reconnaissance Office, 1987.

———. *The Gambit Story.* Center for the Study of National Reconnaissance. Chantilly, VA: National Reconnaissance Office, 1987.

———. *The Hexagon Story.* Center for the Study of National Reconnaissance. Chantilly, VA: National Reconnaissance Office, 1987.

Ogawa, Dennis, and Glen Grant. *Ellison S. Onizuka: A Remembrance.* Kailua-Kona, HI: The Onizuka Memorial Committee, 1986.

Onizuka Alumni Committee. *Onizuka Air Force Station: The Birth and Legacy of Satellite Programs.* Sunnyvale, CA: Privately printed, 2010.

SAFSP Alumni. *Birth of Air Force Satellite Reconnaissance: Facts, Recollections and Reflections.* Los Angeles, CA: SAFSP Alumni Association, 2015.

Temple, L. Parker. *Shades of Gray: National Security and the Evolution of Space Reconnaissance.* Reston, VA: American Institute of Aeronautics and Astronautics, 2005.

INDEX

About the Sunnyvale Historical Society and Museum Association

The Sunnyvale Historical Society and Museum Association (SHSMA) is a nonprofit organization whose mission is to tell the stories of the diverse history of Sunnyvale and its surrounding area through collections, exhibits, and educational programs. The society was founded in 1956 by a number of public-spirited citizens to rescue Bay View ranch, which was the original home of the Martin Murphy family, early settlers in Santa Clara County whose arrival in California predated the Donner Party. Though unable to save the house from demolition, the historical society had a hard lesson reinforced: if there is no active organization to preserve local heritage, it will be gone forever.

The museum contains permanent and rotating exhibits for the public to understand the evolution of the Sunnyvale area, and it educates people of all ages through docent-led presentations for school children and adults, groups from senior centers, members of other museum societies, and other various organizations who are visiting. In addition to providing a unique presentation of local history, another important goal of SHSMA is to collect and preserve artifacts entrusted to its care.

SHSMA has recently shared plans to expand the museum to create a new permanent exhibit on national defense and reconnaissance, with a significant portion dedicated to the story of Onizuka Air Force Station's Blue Cube. This recent history highlights an important part of Sunnyvale history associated with the development of technology and defense. Though the full significance of the top-secret operations that took place at the Blue Cube has yet to be fully revealed, the historical society realizes its obligation as curator of history to collect artifacts and stories from this important time before they have been forgotten and share them with future generations.

Sunnyvale Heritage Park Museum
570 East Remington Drive
Sunnyvale, CA 94087
info@heritageparkmuseum.org
heritageparkmuseum.org

408-749-0220
www.facebook.com/SunnyvaleMuseum

DISCOVER THOUSANDS OF LOCAL HISTORY BOOKS
FEATURING MILLIONS OF VINTAGE IMAGES

Arcadia Publishing, the leading local history publisher in the United States, is committed to making history accessible and meaningful through publishing books that celebrate and preserve the heritage of America's people and places.

Find more books like this at
www.arcadiapublishing.com

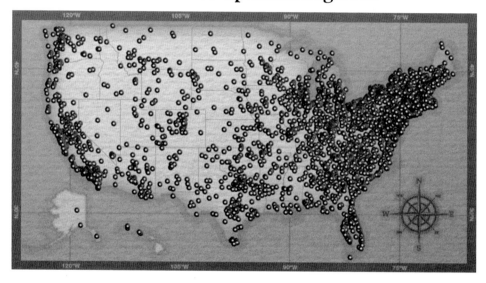

Search for your hometown history, your old stomping grounds, and even your favorite sports team.

Consistent with our mission to preserve history on a local level, this book was printed in South Carolina on American-made paper and manufactured entirely in the United States. Products carrying the accredited Forest Stewardship Council (FSC) label are printed on 100 percent FSC-certified paper.

MADE IN THE